Frithjof Christian Küpper

Biology and biochemistry of host-pathogen interactions in marine brown algae

Hartung-Gorre Verlag, Konstanz

Bibliographic information published by Die Deutsche Nationalbibliothek

Die Deutsche Nationalbibliothek lists this publication in the Deutsche Nationalbibliografie; detailed bibliographic data is available in the internet at http://dnb.ddb.de.

First edition 2001

HARTUNG-GORRE VERLAG KONSTANZ
http://www.hartung-gorre.de

ISBN-10: 3-89649-695-6
ISBN-13: 978-3-89649-695-9

Biology and biochemistry of host-pathogen interactions in marine brown algae

Dissertation
zur Erlangung des Grades eines Doktors der Naturwissenschaften
im Fachbereich Biologie
der mathematisch-naturwissenschaftlichen Sektion
der Universität Konstanz

&

Thèse pour l'obtention du titre de Docteur en Sciences de l'Université
Paris XI Orsay

vorgelegt von

par

Frithjof Christian Küpper

Konstanz & Orsay, 2001

The role of seaweeds and their iodine in traditional life in the Aran Islands (Co. Galway, Ireland) one century ago, as described by J. M Synge

(The Aran Islands. George Allen & Unwin Ltd., London, 1907)

"The people have taken advantage of this dry moment to begin the burning of kelp, and all the islands are lying in a volume of grey smoke. There will not be a very large quantity this year, as the people are discouraged by the uncertainty of the market, and do not care to undertake the task of manufacture without a certainty of profit.

The work needed to form a ton of kelp is considerable. The seaweed is collected from the rocks after the storms of autumn and winter, dried on fine days, and then made up into a rick, where it is left till the beginning of June.

It is then burnt in low kilns on the shore, an affair that takes from twelve to twenty-four hours of continuous hard work, though I understand the people here do not manage well, and spoil a portion of what they produce by burning it more than is required.

The kiln holds about two tons of molten kelp, and when full it is loosely covered with stones, and left to cool. In a few days the substance is as hard as the limestone, and has to be broken with the crowbars before it can be placed in curaghs for transport to Kilronan, where it is tested to determine the amount of iodine it contains, and paid for accordingly. In former years good kelp would bring seven pounds a ton, now four pounds are not always reached.

In Aran even manufacture is of interest. The low flame-edged kiln, sending out dense clouds of creamy smoke, with a band of red and grey clothed workers moving in the haze, and usually some petticoated boys and women who come down with drink, forms a scene with as much variety and colour as any picture from the East."

"This morning we were out again near the pier on the north side of the island. As we paddled slowly with the tide, trolling for pollock, several curaghs, weighed to the gunnel with kelp, passed us on their way to Kilronan."

"The morning had none of the supernatural beauty that comes over the island so often in rainy weather, so we basked in the vague enjoyment of the sunshine, looking down at the wild luxuriance of the vegetation beneath the sea, which contrasts strangely with the nakedness above it."

Contents

1

Introduction

Far from being an insurmountable obstacle to the analysis of an organic system, a pathological disorder is often the key to understanding it.
Konrad Lorenz, 1973

Host-pathogen interactions involve all organisational levels of life: from molecules mediating cross-talk, attack and defense between organisms, to cellular compartments, cells, organisms and finally the trophic levels and fluxes of matter and energy in ecosystems. Consequently, any comprehensive scientific approach has to be interdisciplinary and has to take into account the levels of biological hierarchy involved: biochemistry works alongside with field ecology, cytology and biophysics with biogeography. Host-pathogen interactions are among the best examples of close co-evolution.

Aquatic ecosystems have long been among the favourites of ecological research, yet, this has mainly been restricted to studies on trophic interactions. Marine brown algae and even more their parasites have remained outsiders in this respect. Even though observations of fungi and parasites belonging to other phyla have been made since the foundations of marine botany were laid, their studies remained limited to field observations until very recently. Still, many of the accurate observations made earlier in this and in the last century appear to have been forgotten and received very little interest in recent decades.

In contrast, unravelling the molecular bases of defense in mammals, particularly in humans, has been an unparalleled chain of success in the 19th and 20th century, concomitant with a general rise of immunology and biochemical pharmacology. It started with the discoveries of Louis Pasteur (inducible immunity), Emil von Behring (first mechanistic investigations of immunity, serum antitoxins), Robert Koch (cellular immunity to tuberculosis) and Paul Ehrlich (concept of antigen – antibody recognition). A key element of defense in mammalian cells, the respiratory (oxidative) burst of neutrophile granulocytes, was discovered as early as 1933 by Baldridge and Gerard. The same reaction was discovered in higher plants half a century later (Doke, 1983a, 1983b), even though plants – as sessile, largely immobile organisms – rely even more than animals on inducible defenses. The role of the oxidative burst in the concepts of systemic acquired resistance (Dempsey & Klessig, 1994; Lamb & Dixon, 1997) and the hypersensitive reaction (Levine *et al.*, 1994) are now well-established in plant pathology. Still, until only the last 5 years, it was unknown if the same principles applied also to the marine habitat and to phylogenetic lineages that diverged from the terrestrial ones early in eukaryotic evolution.

Nevertheless, the 19th century had already seen a considerable research activity in pathogens of brown algae, in parallel to the development of marine botany in general. Wright (1879) first described *Eurychasma dicksonii*, which other studies (e.g. Rattray, 1885; Petersen, 1905; Sparrow, 1934) confirmed to be a widespread pathogen of Phaeophyceae. However, all these studies remained limited to field-collected samples. Laboratory studies on brown algal pathogens were not carried out due to lack of adequate culture methods. Physiological and biochemical aspects of the interaction remained unknown, and in most cases, the phylogenetic relationships of the pathogens have not been resolved.

Kelps were at the centre of interest of chemists with the discovery of iodine (Courtois, 1813) in the early 19th century. The biochemistry of iodine in situ found deeper interest only about a century later with the competing studies of Dangeard (1928) and Kylin (1929), who found the release of volatile forms of iodine from seaweeds in parallel. With the availability of radioactive iodine as a tracer, studies of the uptake mechanism became feasible more easily (Baily & Kelly, 1955; Shaw, 1959, 1960). Von Stosch (1962) pointed out the role of iodine as a micronutrient of red algae. Yet, the biological significance of iodine accumulation in marine algae remained mostly unknown.

Although an interest in viruses (e.g. Müller *et al.*, 1998) and endophytes (e.g. Ellertsdóttir & Peters, 1997; Burkhardt and Peters, 1998) infecting brown algae has been awakening in recent years, studies of defense reactions on the physiological and biochemical levels have been lacking. The last five years have also seen considerable progress in red algae with the discoveries of cellular transformation by red algal parasites (Goff & Coleman, 1995), crosstalk mediated by carbohydrate sulfation patterns between a red alga and its pathogen (Bouarab *et al.*, 1999), and of an oligosaccharide-triggered oxidative burst (Weinberger *et al.*, 1999).

My doctoral program was conceived to contribute to a profound, comprehensive understanding of the interactions between brown algae and their pathogens. Most of the experimental work was done with the filamentous form *Pylaiella littoralis* and the kelps *Laminaria digitata* and *Macrocystis pyrifera*.

As a first approach, I investigated populations of the filamentous brown alga, *Pylaiella littoralis*, at various European localities for the presence of eukaryotic pathogens (Chapter 2). I detected frequent epidemics of *Anisolpidium rosenvingei*, *Eurychasma dicksonii* and *Chytridium polysiphoniae. Eurychasma* and *Chytridium* could be established in laboratory cultures for later studies of host-pathogen interactions, including the ecophysiology of their infective behaviour and host range (Chapter 3). A molecular approach clarified their phylogenetic affiliations using sequence homologies of their 18 S rRNA genes (Chapter 4). Since both pathogens are obligate biotrophs, it appeared desirable to study their influence on the energy metabolism of their host. I used a novel technique, fluorescence kinetic microscopy (FKM) which permits spatially resolved measurements of chlorophyll fluorescence induction kinetics at cellular and sub-cellular levels. Their interpretation reveals the state of the photosynthetic apparatus, reflecting the impact of different infection stages of *Eurychasma* and *Chytridium* on the photosynthetic capacity of their host *Pylaiella littoralis* (Chapter 5).

In parallel, I addressed the question whether defense reactions comparable to the oxidative burst in land plants also exist in brown algae. We found that degradation products of alginate can trigger an oxidative burst in brown algae. This reaction implies typical steps of cellular signal transduction such as ion channels, protein phosphorylation and phospholipases (Chapter 6). The accumulation of iodine had long been attributed to various aspects of defense in *Laminaria* species (Wever *et al.*, 1991), though conclusive evidence has been scarce so far. The discovery of vanadium haloperoxidases in brown algae (Vilter *et al.*, 1983; Vilter, 1984) had pointed out the link between the metabolism of iodine and active oxygen species in the apoplast. I undertook a physiological and biochemical approach to study the uptake mechanism and the function of iodine with regard to oxidative stress in kelp plants (Chapter 7). The oxidative burst is followed by a strong iodine efflux. A detailed chemical analysis revealed that it includes iodinated volatile compounds potentially toxic for microorganisms and different inorganic iodine species (Chapter 8).

Finally, the occurrence of the oxidative burst in brown algae and its biological and ecological significance were investigated (Chapter 8). Firstly, a selection of various brown algae were screened for their capacity to respond to elicitation by oligoalginates with an oxidative burst. In a second step, the consequences of elicitation upon attack of *Laminaria digitata* by a pathogenic kelp endopyte, *Laminariocolax tomentosoides*, were addressed. Using axenic *Macrocystis* sporophytes, the role of oligoalginate recognition for maintaining an equilibrium between kelp plants and their bacterial epiflora was explored.

The growing interest in lipid derivatives in stress- and pathogenesis-related signalling in terrestrial plants (e.g. Bostock *et al*., 1992; Vollenweider *et al*., 2000) suggested to investigate the role of bacterial endotoxins, polyunsaturated fatty acids and cyclopentenones in *Laminaria* and *Macrocystis*, regarding the oxidative burst and defense against pathogenic endophytes (Chapter 9).

References

Baily NA, Kelly S, 1955: Iodine exchange in *Ascophyllum*.- Biol. Bull. Woods Hole 109, 13-20

Baldridge CW, Gerard RW, 1933: The extra respiration of phagocytosis.- American Journal of Physiology 103, 235-6

Bostock RM, Yamamoto H, Choi D, Ricker KE, Ward BL, 1992: Rapid stimulation of 5-lipoxygenase activity in potato by the fungal elicitor arachidonic acid.- Plant Physiology 100, 1448-56

Bouarab K, Potin P, J. Correa, and B. Kloareg. 1999. Sulfated oligosaccharides mediate the interaction between a marine red alga and its green algal pathogenic endophyte.- Plant Cell 11, 1635-50

Burkhardt E, Peters AF, 1998: Molecular evidence from nrDNA IST sequences that *Laminariocolax* (Phaeophyceae, Ectocarpales sensu lato) is a worldwide clade of closely related kelp endophytes.- J. Phycol. 34, 682-91

Courtois M, 1813: Sur un nouvel acide formé avec la substance découverte par M. Courtois.- Ann. Chim. Pys. (Paris) 88, 311-8

Dangeard P, 1928: Sur le dégagement de l'iode chez les algues marines.- Comptes Rendus Hebdomadaires des Séances de l'Académie des Sciences 186, 892-4

Dempsey DA, Klessig DF, 1994: Salicylic acid, active oxygen species and systemic acquired resistance in plants.- Trends in Cell Biology 4, 334-8

Doke N, 1983a: Involvement of superoxide anion generation in the hypersensitive response of potato tuber tissues to infection with an incompatible race of Phytophthora infestans and to the hyphal wall components.- Physiological Plant Pathology 23, 345-57

Doke N, 1983b: Generation of superoxide anion by potato tuber protoplasts during the hypersensitive response to hyphal wall components of *Phytophthora infestans* and specific inhibition of the reaction by suppressors of hypersensitivity.- Physiological Plant Pathology 23, 359-67

Ellertsdóttir E, Peters AF, 1997: High prevalence of infection by endophytic brown algae in populations of *Laminaria* spp. (Phaeophyceae).- Marine Ecology Progress Series 146, 135-43

Goff LJ, Coleman AW, 1995: Fate of parasite and host organelle DNA during cellular transformation of red algae by their parasites.- Plant Cell 7, 1899-911

Kylin H, 1929: Über das Vorkommen von Jodiden, Bromiden und Jodidoxydasen bei Meeresalgen.- Hoppe-Seyler's Zeitschrift für Physiologische Chemie 186, 50-84

Lamb C, Dixon RA, 1997: The oxidative burst in plant disease resistance.- Annual Review of Plant Physiology and Plant Molecular Biology 48, 251-75

Levine A, Tenhaken R, Dixon R, Lamb C, 1994: H_2O_2 from the oxidative burst orchestrates the plant hypersensitive disease resistance response.- Cell 79, 583-93

Müller DG, Kapp M, Knippers R, 1998: Viruses in marine brown algae.- Advances in Virus Research 50, 49-67

Petersen HE, 1905: Contributions à la connaissance des phycomycètes marins (Chytridinae Fischer). Kgl. Danske Videnskabernes Selskabs Forhandlinger 5, 439-88

Rattray J, 1885: Note on *Ectocarpus*.- Transactions of the Royal Society of Edinburgh 32, 589-602

Shaw TI, 1959: The mechanism of iodine accumulation by the brown sea weed *Laminaria digitata*. The uptake of I^{131}.- Proceedings of the Royal Society London B 150, 356-71

Shaw TI, 1960: The mechanism of iodine accumulation by the brown sea weed *Laminaria digitata*. Respiration and iodide uptake.- Proceedings of the Royal Society London B 152, 109-17

Sparrow FK Jr., 1934: Observations on marine phycomycetes collected in Denmark.- Dansk Botanisk Arkiv 8(6), 1-24

Vilter H, Glombitza KW, Grawe A (1983) Peroxidases from Phaeophyceae I: extraction and detection of the peroxidases. Botanica Marina 26, 331-40

Vilter H (1984) Peroxidases from Phaeophyceae: A vanadium (V)-dependent peroxidase from *Ascophyllum nodosum*. Phytochemistry 23, 1387-90

Vollenweider S, Weber H, Stolz S, Chételat A, Farmer EE, 2000: Fatty acid ketodienes and fatty acid ketotrienes: Michael addition acceptors that accumulate in wounded and diseased *Arabidopsis* leaves.- Plant Journal 24 (4), 467-76

von Stosch HA, 1962: Jodbedarf bei Meeresalgen.- Naturwissenschaften 49, 42-3

Weinberger F, Friedlander M, Hoppe H-G, 1999: Oligoagars elicit an oxidative burst in *Gracilaria conferta* (Rhodophyta).- Journal of Phycology 35, 747-55

Wever R, Tromp MGM, Krenn BE, Marjani A, Van Tol M, 1991: Brominating activity of the seaweed *Ascophyllum nodosum*: impact on the biosphere.- Environ. Science & Technol. 25, 446-9

Wright EP, 1879: On a species of *Rhizophydium* parasitic on species of *Ectocarpus*, with notes on the fructification of the *Ectocarpi*. Transactions of the Royal Irish Academy 26, 369-79

2

Nova Hedwigia **69** 3—4 381—389 Stuttgart, November 1999

Massive occurrence of the heterokont and fungal parasites *Anisolpidium, Eurychasma* and *Chytridium* in *Pylaiella littoralis* (Ectocarpales, Phaeophyceae)

by

Frithjof C. Küpper*) and Dieter G. Müller

Universität Konstanz, Fakultät für Biologie, PO Box 5560, D-78457 Konstanz, Germany
E-mail: frithjof.kuepper@uni-konstanz.de

With 6 figures

Küpper, F.C. & D.G. Müller (1999): Massive occurrence of the heterokont and fungal parasites *Anisolpidium, Eurychasma* and *Chytridium* in *Pylaiella littoralis* (Ectocarpales, Phaeophyceae). - Nova Hedwigia 69: 381-389.

Abstract: The abundance of fungal and heterokont parasites in the marine brown alga *Pylaiella littoralis* has been followed for three years in a tidal flat at Roscoff (Brittany). Additional observations were made in Shetland and Ireland. In Roscoff, an epidemic of *Anisolpidium rosenvingei* (Anisolpidiales, Oomycota, Heterokonta) occurred every late autumn, coinciding with the typical decline of *Pylaiella* in winter. Lower densities of *Anisolpidium* were found in Roscoff and on the Irish coast in early summer. In Shetland and Ireland, *Pylaiella* was found with massive infections by two other parasites: *Eurychasma dicksonii* (Saprolegniales, Oomycota, Heterokonta) and *Chytridium polysiphoniae* (Chytridiomycota, Eumycota). Whilst *E. dicksonii* and *C. polysiphoniae* can infect all cells of *P. littoralis*, *A. rosenvingei* specifically infects sporangia of *Pylaiella*, which results in a reduction of the host's reproductive potential.

Kurzfassung: Der Befall der marinen Braunalge *Pylaiella littoralis* durch Pilze und heterokonte Parasiten wurde im Gezeitenbereich von Roscoff (Bretagne) über drei Jahre beobachtet. Weitere Einzelbeobachtungen erfolgten in Shetland und Irland. In Roscoff trat jeden Spätherbst eine Epidemie von *Anisolpidium rosenvingei* (Anisolpidiales, Oomycota, Heterokonta) auf, zusammen mit dem typischen Rückgang von *Pylaiella* im Winter. In geringerer Dichte trat *Anisolpidium* auch im Frühsommer in Roscoff und an der irischen Küste auf. In Shetland und Irland wurden massive Infektionen auf *Pylaiella* durch zwei andere Parasiten beobachtet: *Eurychasma dicksonii* (Saprolegniales, Oomycota, Heterokonta) und *Chytridium polysiphoniae* (Chytridiomycota, Eumycota). Während *Eurychasma* und *Chytridium* alle Zellen eines *Pylaiella*-Filaments befallen können, befällt *Anisolpidium* ausschließlich Sporangien, was zu einer deutlichen Verringerung des Fortpflanzungspotentials des Wirtes führt.

Key words: *Anisolpidium rosenvingei*, *Chytridium polysiphoniae*, epidemiology, *Eurychasma dicksonii*, Heterokonta, Phaeophyceae, *Pylaiella littoralis*, parasitism.

*)To whom correspondence should be addressed.
Current address: Station Biologique, CNRS UMR 1931, Place G. Teissier, BP74, F-29682 Roscoff, Brittany, France.

0029-5035/99/0069-0381 $ 2.25

Introduction

Fungal infections in brown algae are known for a long time (Andrews 1976, Correa 1997), but knowledge of their epidemiology and life cycles is still very scarce. During a search for pathogens in natural populations of *Pylaiella littoralis* (L.) Kjellman, we encountered massive prevalences of parasitic fungi and protists.

Pylaiella littoralis is a cosmopolitan, filamentous brown alga classified in the order Ectocarpales. Parasitism has been reported repeatedly in *Pylaiella* and related genera in the past. Some of these pathogens are now recognised to belong to the lineage of heterokonts (Saunders et al. 1997). Three eukaryotic parasites are presently known to infect *Pylaiella*: 1. *Eurychasma dicksonii* (Wright) Magnus (Saprolegniales, Oomycota; Wright 1879, Dangeard 1934, Feldmann 1954, Sparrow 1960, Jenneborg 1977, Wilce et al. 1982, Konno & Tanaka 1988, Siemer & Pedersen 1995). 2. *Anisolpidium rosenvingei* (H.E. Petersen) Karling (Petersen 1905, Karling 1943, Sparrow 1960). This taxon was placed in the Hyphochytriales, Hyphochytriomycota by Sparrow (1960), and more recently in the Anisolpidiales, Oomycota by Hawksworth et al. (1996). 3. *Chytridium polysiphoniae* Cohn (Chytridiales, Eumycota; Cohn 1865, Petersen 1905, Feldmann 1954, Sparrow 1960).

P. littoralis populations exhibit marked seasonal fluctuations in the intertidal zone (Cardinal 1964, Steele & Whittick 1991, Russell 1963). At the Breton coast it is a summer annual, with vegetative growth starting in March. Wilce et al. (1982) correlated seasonal abundance of free-floating *Pylaiella littoralis* on the coast of Massachusetts with the presence of *Eurychasma dicksonii*.

The aim of this study was to verify previous observations on the presence and abundance of eukaryotic pathogens in *P. littoralis* on European coasts. We report here observations on fungal and heterokont parasites and seasonal fluctuations of *P. littoralis* in Roscoff (Brittany) and localities in Shetland and Ireland.

Materials and methods

Sampling sites for field material were the tidal flat in front of the Biological Station at Roscoff (Brittany; September 1995 to November 1997), the east pier of the harbour of Howth (Co. Dublin, Ireland; June 1996), Dalkey (Killiney Bay, Co. Dublin, Ireland; June 1996), Brei Wick (Lerwick, Shetland; April/May and July 1996), Aith Voe and Noss Sound (Bressay, Shetland; April/May and July 1996; Fig. 1). 50 to 100 tufts of *P. littoralis* (average blotted fresh weight 0.1 g) were collected haphazardly each time at a given site. In total, more than 3400 samples were screened. In Roscoff, regular observations were carried out for 26 months, with intervals of four weeks or less during the maximum period of algal development. The material was examined immediately after collection with a compound light microscope at 100× magnification. We recorded the reproductive status of *Pylaiella* and the prevalence of infections. A specimen was considered fertile or infected, respectively, if at least one sporangium or infection was observed. Field biomass of *P. littoralis* at Roscoff was roughly estimated by thallus size and the time required to collect a given number of specimens.

Surface seawater temperatures for Roscoff, which had been obtained in 1 km distance from the sampling site at Astan, were taken from Conq et al. (1998).

For permanent mounts, *P. littoralis* filaments were fixed and stained in a drop of acetocarmine solution on a microscopic slide and mounted in 50% Karo® syrup.

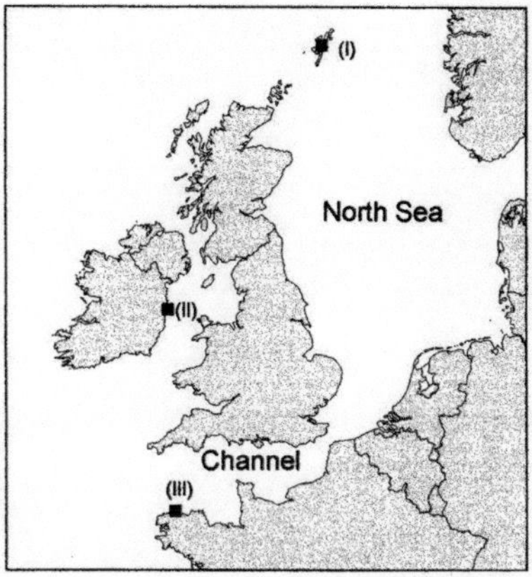

Fig. 1. Sampling sites of *Pylaiella littoralis* on the European Atlantic coast. (I) Shetland; (II) Howth, Co. Dublin, Ireland; (III) Roscoff, Brittany.

Results

At Roscoff, the parasite which caused an epidemic in *Pylaiella littoralis* in late 1995, 1996 and again in 1997 was identified as *Anisolpidium rosenvingei* (Fig. 2). The same species was present in *P. littoralis* at Howth, together with *Eurychasma dicksonii* (Fig. 3) in June 1996. The two parasites found in *P. littoralis* in Shetland in April and July 1996 were identified as *E. dicksonii* and *Chytridium polysiphoniae* (Fig. 4). In several cases, the latter were found together on the same host filaments. *E. dicksonii* was found on *Pylaiella* in Shetland and Ireland in June and July 1996. These three parasites followed different infection patterns: *A. rosenvingei* affected exclusively the reproductive structures of the host (Fig. 2), i.e. it was found only in unilocular and plurilocular sporangia of *Pylaiella*. *E. dicksonii* and *C. polysiphoniae* infected both vegetative cells and sporangia of *Pylaiella*. All three parasite species are holocarpic. Whilst *Anisolpidium* and *Eurychasma* formed their sporangia within the host cell (Figs 2, 3), which in the case of *Eurychasma* became considerably enlarged, sporangia of *Chytridium* were produced on the surface of the host thallus (Fig. 4), whose cell wall was penetrated by a haustorial structure (Fig. 5).

At Roscoff, *A. rosenvingei* occurred as an epidemic in up to 70% of the *P. littoralis* population by mid-October in the years 95-97 (Fig. 6). This coincided with maximum fertility of *P. littoralis* and the onset of its decline towards the cold season. Both

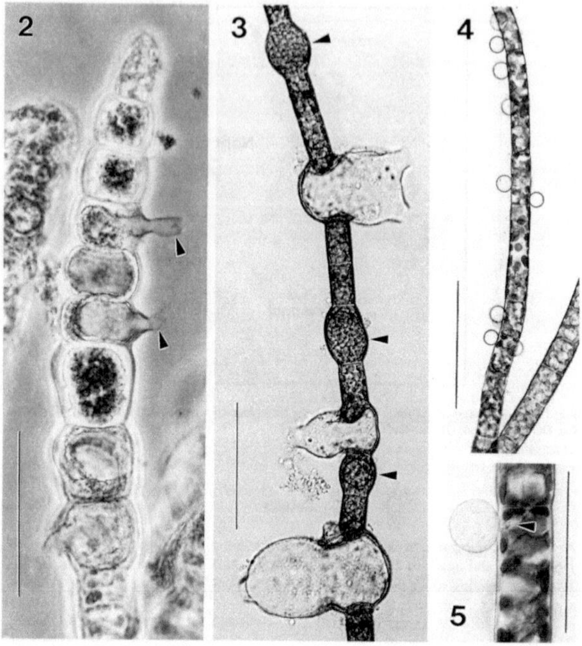

Figs 2-5. Parasites in *Pylaiella littoralis*. – Fig. 2. *Anisolpidium rosenvingei*, fixed material from Roscoff. All cells in a series of *Pylaiella* sporangia are infected by different stages of *Anisolpidium*, including empty sporangia with lateral exit tubes (arrowheads). Phase contrast, scale bar 50 mm. – Fig. 3. *Eurychasma dicksonii*. Different stages (arrowheads) and empty sporangia of the parasite. Scale bar 100 mm. – Figs 4, 5. *Chytridium polysiphoniae*. Fig. 4. High infection density on living host, scale bar 100 mm. Fig. 5. Sporangium with haustorial structure (arrowhead) in host cytoplasm. Scale bar 50 mm.

species disappeared almost totally by mid-December. Between January and late March, *P. littoralis* was virtually absent, and an intensive search on neighbouring areas of the coast yielded only very small numbers of short filaments, some of which were fertile. By early April, small *Pylaiella* thalli reappeared, often heavily covered by

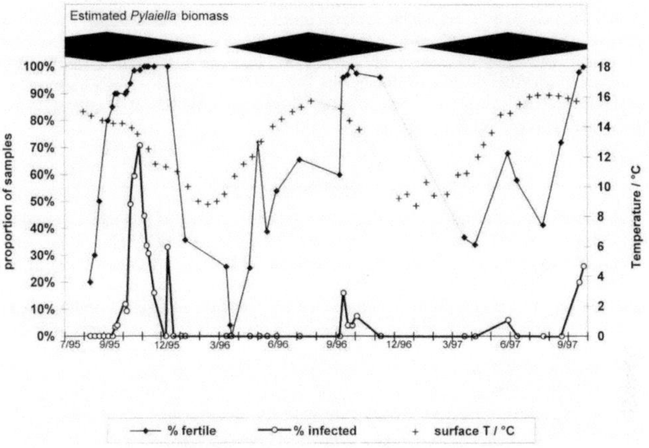

Fig. 6. Host fertility and abundance of infections by *A. rosenvingei* in *P. littoralis*, and seawater temperature at Roscoff (Brittany) 1995-97. Estimation of *Pylaiella* biomass as described in Materials & Methods.

diatoms. In April and May, the population and individual size of *P. littoralis* increased strongly, and the density of epiphytic diatoms decreased. Maximum *Pylaiella* biomass was reached in June-July. Fertility of *P. littoralis* increased from 0-38% in April-May to a first maximum of ca. 70% in June. During this period, *A. rosenvingei* was absent in 1996 and its prevalence was below 7% in 1997. During maximum abundance, most *Pylaiella* at Roscoff was found epiphytically on *Fucus serratus* L., and in smaller numbers on *F. vesiculosus* L. and *Ascophyllum nodosum* (L.) Le Jolis. Complete fertility of *Pylaiella* (100%) occurred in autumn (consistent with the observations by Feldmann 1954), in synchrony with the highest prevalence of *Anisolpidium*. However, maximum values differed markedly between 72% in 1996, and 24% in 1997.

In Howth, Ireland, cumulative prevalence of infection by *Anisolpidium* and *Eurychasma* was 10%, whereas the value for *Eurychasma* and *Chytridium* in Aith Voe in Shetland was 10.3%.

Maximum outbreaks of *A. rosenvingei* occurred at water temperatures between 11 and 13°C (a minor epidemic in 1996 at >14°C), while infections by *E. dicksonii* and *C. polysiphoniae* were observed at water temperatures between 6 and 13°C.

In none of the over 3400 samples of *P. littoralis* examined by light microscopy symptoms of virus infection (Markey 1974, Maier et al. 1998) were found.

Voucher specimens (permanent mounts on microscope slides) of *Anisolpidium rosenvingei* infecting *Pylaiella littoralis* are deposited in the following collections: University of California at Berkeley (UC 1714811), Farlow Herbarium, Cambridge, Massachusetts (FH), Station Biologique de Roscoff (Herbarium J. Cabioch), Muséum National d'Histoire Naturelle, Paris (PC), British Museum/Natural History, London (BM) and Rijksherbarium, Leiden (L).

Discussion

The general aim of our study was to assess the occurrence of pathogens in European populations of *Pylaiella littoralis*. We did not observe any symptoms of viral infections, in contrast to North American Atlantic and Pacific coasts (Markey 1974, Maier et al. 1998). Instead, infections by *A. rosenvingei*, *E. dicksonii*, and *C. polysiphoniae* were notorious. We found all three parasites as regular components of the marine flora at the examined locations. They follow different infection strategies and belong to different phylogenetic lineages: the Oomycota are heterokonts like their brown algal host, while the Chytridiomycota are considered genuine fungi (Barr 1992). Our records of *A. rosenvingei* are the first for Brittany and Ireland. So far, the species has only been reported from the European side of the North Atlantic: Petersen (1905) observed it in Denmark and Greenland, Sparrow (1934) reported it from Denmark, and Aleem (1955) found it in Sweden. Our findings of *E. dicksonii* and *C. polysiphoniae* are first records for Shetland, yet, they were not observed in Roscoff throughout the whole study despite a previous report by Dangeard (1934). *A. rosenvingei*, *C. polysiphoniae*, and *E. dicksonii* are the only known eukaryotic pathogens of *P. littoralis*. A report of *Phagomyxa algarum* Karling (Plasmodiophoromycota) in vegetative cells of *Pylaiella fulvescens* Bornet by Karling (1944) does not refer to *Pylaiella*, since the host is now known as *Bachelotia antillarum* (Grunow) Gerloff.

Whilst *A. rosenvingei* infects only sporangia of the host, *E. dicksonii* and *C. polysiphoniae* infect mainly younger vegetative cells of *Pylaiella* filaments. The consequences of infection by *E. dicksonii* are different from those by *A. rosenvingei* and *C. polysiphoniae*. Once the spores of the parasite have been released, the empty *Eurychasma* sporangium causes fragmentation of the *Pylaiella* filaments. According to Wilce et al. (1982), this constitutes an effective means of vegetative propagation that may contribute to the buildup of brown tides. Jenneborg (1977) recognised previous descriptions of variable unilocular sporangia in *P. varia* Kjellman as infections by *E. dicksonii* and relegated the species into *P. littoralis*. Siemer & Pedersen (1995) and Siemer (1998), however, provided evidence from culture and molecular studies that *P. varia* is different from *P. littoralis*.

The specific infection of sporangia of *P. littoralis* by *Anisolpidium* has not been reported previously. It is likely to interfere with the host's reproductive potential. This possibility was examined by correlating *Anisolpidium* prevalence with abundance and fertility of the host at Roscoff. Here, seasonal prevalence of *Anisolpidium* appears to contribute to the decline of the *Pylaiella* population in autumn (Feldmann 1954). Observation of this event in three consecutive years justifies the conclusion that it is a typical seasonal pattern not recognized previously. However, the weaker epidemic in autumn 1996 indicates that additional, unknown factors may contribute to the seasonal decline of *P. littoralis* at Roscoff. Fig. 6 shows a clear correlation of *Anisolpidium* prevalence in *Pylaiella* with the reproductive periods of *Pylaiella* during rather short times of the year. The short reproductive period of *Pylaiella* may be a response to the infective pressure by *A. rosenvingei*, since parasite epidemics would be enhanced by a more continuous availability of the host's sporangia. The interaction between *Anisolpidium* and *Pylaiella* suggests an analogy to a predator-prey-interaction, which are also known to create population density cycles (Volterra 1926, Lotka 1932, MacLulick 1937).

A. rosenvingei was found in *Pylaiella* only during short periods of the year (Fig. 6) when offshore surface seawater temperatures ranged between 11 and 16°C. It remains unclear how it survives for the rest of the year. Johnson (1957) observed thick-walled spores of *A. ectocarpii* inside the cells of *Hincksia mitchelliae* (Harvey) Silva. He concluded that they are resting stages enabling the parasite to survive unfavourable conditions.

The effect of temperature as a major abiotic factor controlling both host and parasite abundance is likely to be rather complex. Our observations were made in intertidal zones which were exposed for at least 3 h per tidal cycle and where temperature fluctuations must be expected to deviate significantly from the rather constant sea surface temperatures measured offshore. Furthermore, our observations on the abundance of *E. dicksonii* and *C. polysiphoniae* were not carried out continuously. The compilations by Sparrow (1960) and Jenneborg (1977) suggest that the tolerance of *A. rosenvingei* and *C. polysiphoniae* to temperature and salinity differences is narrower than that of *E. dicksonii*. They are restricted to the cold-temperate and polar regions of the North Atlantic, and they occur predominantly in regions with maximum temperatures between 5 and 13°C. The lack of records from the Baltic and the Mediterranean Sea may reflect their low tolerance towards low or high salinity levels, respectively, and higher temperatures.

All present knowledge on eukaryotic pathogens of filamentous brown algae is based on field observations. The establishment of laboratory cultures can be expected to contribute significant progress in various areas such as the host range of parasites, effects of simultaneous infection by more than one pathogen, and confirmation of their correct taxonomic position with molecular methods.

Acknowledgements

We express our thanks to S. Loiseaux-de Goër and B. Kloareg (Roscoff) for discussions, laboratory space and travelling funds for F.C.K. Scholarships to F.C.K. were provided by Studienstiftung des Deutschen Volkes (Bonn) and Hüls AG-Stiftung (Marl). I. Maier (Konstanz) is thanked for a critical reading of the manuscript.

References

ALEEM, A.A. (1955): Marine fungi from the west-coast of Sweden. - Ark. Bot. **3**: 1-35.

ANDREWS, J.H. (1976): The pathology of marine algae. - Biol. Rev. **51**: 211-253.

BARR, D.J.S. (1992): Evolution and kingdoms of organisms from the perspective of a mycologist. - Mycologia **84**: 1-11.

CARDINAL, A. (1964): Etude sur les Ectocarpacées de la Manche. - Beih. Nova Hedwigia **15**: 1-86, 41 figs.

CONQ, C., J.-C DAUVIN, J.-C. LORGERÉ & Service Mer de la Station Biologique de Roscoff (1998): Température et salinité de l'eau de mer au large de Roscoff de 1991 à 1997. - Cah. Biol. Mar. **39**: 207-212.

COHN, F. (1865): Chytridii species novae marinae. - Hedwigia **4**: 169-170.

CORREA, J.A. (1997): Infectious diseases of marine algae: current knowledge and approaches. - Progr. Phycol. Res. **12**: 149-180.

DANGEARD, P. (1934): Sur la présence à Roscoff d'une Chytridiale parasite des Ectocarpacées, l'*Eurychasma dicksonii* (Wright) Magnus. - Ann. Protistol. **4**: 69-73.

FELDMANN, J. (1954): Inventaire de la Flore Marine de Roscoff: Algues, champignons, lichens et spermatophytes. - Trav. Stat. Biol. Roscoff, Suppl. **6**, Roscoff.

HAWKSWORTH, D.L., P.M. KIRK, B.C. SUTTON & D.N. PEGLER (1996): Ainsworth & Bisby's dictionary of the fungi. - 8th ed., CAB International, New York, Oxon.

JENNEBORG, L.-H. (1977): *Eurychasma*-infection of marine algae. Changes in algal morphology and taxonomical consequences. - Bot. Mar. **20**: 499-507.

JOHNSON, T.W. (1957): Resting spore development in the marine phycomycete *Anisolpidium ectocarpii*. - Am. J. Bot. **44**: 875-878.

KARLING, J.S. (1943): The life history of *Anisolpidium ectocarpii* gen.nov. et sp. nov., and a synopsis and classification of other fungi with anteriorly uniflagellate zoospores. - Am. J. Bot. **30**: 637-648.

KARLING, J.S. (1944): *Phagomyxa algarum* n. gen., n. sp., an unusual parasite with plasmodiophoralean and protomyxean characteristics. - Am. J. Bot. **31**: 38-52.

KONNO, K. & J. TANAKA (1988): *Eurychasma dicksonii*, a parasitic fungus on *Acinetospora crinita* (brown alga) in Japan. - Bull. Natl. Sci. Mus., Tokyo, Series B, **14**: 119-122.

LOTKA, A.J. (1932): The growth of mixed populations: two species competing for a common food supply. - J. Washington Acad. Sci. **24**: 199-216.

MACLULICK, D.A. (1937): Fluctuations in numbers of the varying hare (*Lepus americanus*).- University of Toronto Studies, Biology Series **43**: 1-136.

MAIER, I., S. WOLF, N. DELAROQUE, D.G. MÜLLER & H. KAWAI (1998): A DNA virus infecting the marine brown alga *Pilayella littoralis* (Ectocarpales, Phaeophyceae) in culture. - Eur. J. Phycol. **33**: 213-220.

MARKEY, D.R. (1974): A possible virus infection in the brown alga *Pilayella littoralis*. - Protoplasma **80**: 223-232.

PETERSEN, H.E. (1905): Contributions à la connaissance des phycomycètes marins (Chytridinae Fischer). - Overs. Kongel. Danske Vidensk. Selsk. Forh. Medlemmers Arbeider **5**: 439-488.

RUSSELL, G. (1963): A study in populations of *Pilayella littoralis*. - J. Mar. Biol. Assoc. U. K. **43**: 469-483.

SAUNDERS G.W., D. POTTER & R.A. ANDERSEN (1997): Phylogenetic affinities of the Sarcinochrysidales and Chrysomeridales (Heterokonta) based on analyses of molecular and combined data. - J. Phycol. **33**: 310-318.

SIEMER, B.L. (1998): Molecular phylogenetic studies in the simple brown algae. - Ph.D. Thesis, University of Copenhagen.

SIEMER, B.L. & P.M. PEDERSEN (1995): The taxonomic status of *Pilayella littoralis*, *P. varia* and *P. macrocarpa* (Pilayellaceae, Fucophyceae). - Phycologia **34**: 257-266.

SPARROW, F.K. Jr. (1960): Aquatic Phycomycetes. - 2nd revised edition, The University of Michigan Press, Ann Arbor.

SPARROW, F.K. Jr. (1934): Observations on marine phycomycetes collected in Denmark. - Dansk Botanisk Arkiv **8**(6): 1-33.

STEELE, D.H. & A. WHITTICK (1991): Seasonal variation in *Pilayella littoralis* (Phaeophyceae) and its consequences as a food source for the amphipod, *Gammarus lawrencianus*, in the intertidal of Newfoundland. - J. Mar. Biol. Assoc. U. K. **71**: 883-889.

VOLTERRA, V. (1926): Variations and fluctuations of the numbers of individuals in animal species living together. Reprinted in 1931. In: Chapman, R.N. (ed.): Animal Ecology. McGraw Hill, New York.

WILCE, R.T., C.W. SCHNEIDER, A.V. QUINLAN & K. VAN DEN BOSCH (1982): The life history and morphology of free-living *Pilayella littoralis* (L.) Kjellman (Ectocarpaceae, Ectocarpales) in Nahant Bay, Massachusetts. - Phycologia **21**: 336-354.

WRIGHT, E.P. (1879): On a species of *Rhizophydium* parasitic on species of *Ectocarpus*, with notes on the fructification of the *Ectocarpi*. - Trans. Roy. Irish Acad. **26**: 369-379.

Received 6 April 1998, accepted in revised form 17 March 1999.

3

Phycological Research 1999; **47**: 217–223

Infection experiments reveal broad host ranges of *Eurychasma dicksonii* (Oomycota) and *Chytridium polysiphoniae* (Chytridiomycota), two eukaryotic parasites in marine brown algae (Phaeophyceae)

D. G. Müller,[1]* F. C. Küpper[2] and H. Küpper[1]
[1]Faculty of Biology, University of Konstanz, D-78457 Konstanz, Germany and [2]Biological Station, CNRS UPR 9042, BP74, F-29682 Roscoff, France.

SUMMARY

Unialgal cultures of the brown alga *Pylaiella littoralis* (L.) Kjellman infected by either *Eurychasma dicksonii* (Wright) Magnus (Oomycota) or *Chytridium polysiphoniae* (Cohn) H. E. Petersen (Chytridiomycota) were used to elaborate the host ranges of these pathogens. Infection experiments with 48 host species covering 13 orders of the Phaeophyceae showed that 45 species were susceptible to attack by *Eurychasma* and 23 to *Chytridium*. The two pathogens showed host-specific differences in generation times: while in *Pylaiella* the shortest cycles were 16 days for *Eurychasma* and 6 days for *Chytridium*, one and five days more, respectively, were required in *Acinetospora*. Heavy parasite attack on the microscopic stages of host species with heteromorphic life histories, like kelps (Laminariales), is documented and discussed as a potential regulatory factor for the population dynamics of macroalgae.

Key words: Chytridiomycota, *Chytridium*, *Eurychasma*, host range, Oomycota, parasitism, Phaeophyceae, *Pylaiella*, temperature tolerance.

INTRODUCTION

Marine brown algae have long been known to be hosts for viruses and various eukaryotic pathogens. While viruses and endophytes are well covered by recent experimental studies (Kawai and Tokuyama 1995; Peters and Ellertsdóttir 1996; Müller *et al.* 1998), no laboratory studies with fungal and other eukaryotic parasites have been reported.

Pylaiella littoralis (L.) Kjellman has been frequently described with infections by *Eurychasma dicksonii* (Wright) Magnus, which is classified with the Saprolegniales in the Oomycetes (Dangeard 1934; Feldmann 1954; Sparrow 1960; Jenneborg 1977; Wilce *et al.* 1982; Konno and Tanaka 1988; Siemer and Pedersen 1995). Other parasites of *Pylaiella* are the Hyphochytriomycete *Anisolpidium rosenvingii* (H. E. Petersen) Karling (Petersen 1905; Karling 1943; Sparrow 1960; Küpper and Müller 1999) and the Chytridiomycete *Chytridium polysiphoniae* (Cohn) H. E. Petersen (Petersen 1905; Feldmann 1954; Sparrow 1960).

The spelling of the genus name *Pylaiella* has been the subject of controversy over the years and we have followed here the most recent version recommended by Silva *et al.* (1999).

Numerous reports in the literature, and the compilations by Sparrow (1960) and Jenneborg (1977) document that *C. polysiphoniae* and *E. dicksonii* have a broad host range comprising many brown and even some red algae. As all these reports describe infection structures found in field-collected material, it cannot be decided whether infection symptoms in different host species are caused by one parasite species with a very broad host range, or by a multitude of morphologically similar pathogenic taxa with narrow host ranges. In order to clarify this situation, we established defined cocultures of *P. littoralis* with *C. polysiphoniae* and *E. dicksonii* and used them to examine the parasites' competence to infect clonal cultures of selected host species representing many orders of the brown algae. We also report observations on the temperature tolerance of these pathogens.

MATERIALS AND METHODS

Field specimens of *P. littoralis* infected with *Chytridium* and *Eurychasma* were collected at Aith Voe (Bressay, Shetland) in April 1996 as described by Küpper and Müller (1999). The morphological features of the pathogens (holocarpic character, sporangium morphology and size) corresponded well with the descriptions given by Sparrow (1960) and Karling (1977) for *C. polysiphoniae* and by Jenneborg (1977) for *E. dicksonii*. Small tufts of infected *Pylaiella* were coincubated

*To whom correspondence should be addressed.
Email: <Dieter.Gerhard.Mueller@uni-konstanz.de>
Communicating editor: H. Kawai
Received 6 February 1999; accepted 21 May 1999.

Table 1. Infection experiments with the parasites *Eurychasma dicksonii* and *Chytridium polysiphoniae* from *Pylaiella littoralis* to other brown algal hosts

Recipient	Life phase, origin	*Eurychasma*	*Chytridium*
Ectocarpales			
Acinetospora crinita (Carmichael) Kornmann	New Zealand	+	+
	Australia	+	+
Asteronema rhodochortonoides (Boergesen) Müller et Parodi	Canary Islands	+	– (2)
Botrytella uvaeformis (Pringsheim) Kornmann et Sahling	Helgoland (Germany)	+	+
Botrytella micromora Bory	British Channel	+	– (2)
Ectocarpus siliculosus (Dillwyn) Lyngbye	g, New Zealand	+	+
	sp	+	+
Ectocarpus fasciculatus Harvey	Ireland	+	+
	Brittany	– (3)	+
Feldmannia globifer (Kuetzing) Hamel	Ireland	– (2)	+
Feldmannia irregularis (Kuetzing) Hamel	Chile	+	+
	Canary Islands	+	– (3)
Feldmannia simplex (Crouan) Hamel	Ireland	+	+
Hincksia hincksiae (Harvey) Silva	Brittany	+	+
Hincksia mitchelliae (Harvey) Silva	South Africa	+	+
	Canary Islands	+	+
	Argentina	+	– (2)
	North Carolina	– (2)	– (2)
	Japan	– (2)	– (2)
Hincksia sandriana (Zanardini) Silva	Chile	+	– (2)
Kuckuckia spec.	Norway	+	– (2)
	Yugoslavia	+	– (2)
Kuetzingiella battersii (Bornet) Kornmann	Canary Islands	+	– (2)
Pilayella littoralis (L.) Kjellman	Chile	+	+
	Bering Sea	+	+
Spongonema tomentosum (Hudson) Kuetzing	Canary Islands	+	+
Dictyosiphonales			
Cladothele decaisne Hooker et Harvey	Patagonia	+	– (3)
Desmotrichum undulatum (J. Agardh) Reinke	Bering Sea	+	– (2)
Dictyosiphon foeniculaceus Greville	Sweden	+	– (2)
Giraudya sphacelarioides Derbes et Solier	Canary Islands	+	– (2)
Litosiphon pusillus (Carmichael ex Hooker) Harvey	Helgoland (Germany)	+	– (2)
Myriotrichia clavaeformis Harvey	Argentina	+	+
Sphacelariales			
Halopteris paniculata (Suhr) Prud'homme van Reine	New Zealand	+	– (2)
Halopteris sp.	Australia	+	+
Sphacelaria rigidula Kuetzing	California	+	– (3)
Tilopteridales			
Haplospora globosa Kjellman	Helgoland (Germany)	– (2)	+
Tilopteris mertensii (Turner in Smith)	Helgoland (Germany)	+	– (2)
Cutleriales			
Cutleria multifida (Smith) Greville	France, Mediterranean Sea	+	– (2)
Syringodermatales			
Syringoderma phinneyi Henry et Müller	British Columbia, Canada	+	– (3)
Chordariales (microthalli)			
Elachista stellaris Areschoug	France, Mediterranean Sea	+	– (2)
Eudesme virescens (Carmichael ex Harvey in Hooker) J. Agardh	Ireland	+	+
Myriogloea chilensis (Montagne) Llaña	Chile	+	+
Stilophora rhizodes J. Agardh	Sweden	+	– (2)
Scytothamnales			
Scytothamnus fasciculatus (Hooker et Harvey) Cotton	Australia	+	– (2)
Splachnidium rugosum (L.) Greville	Tasmania	+	– (3)
Sporochnales			
Carpomitra costata (Stackhouse) Batters	g, France, Mediterranean Sea	+	+
Perithalia caudata (Labillardière) Womersley	g, Tasmania	+	+
Sporochnus pedunculatus (Hudson) C. Agardh	g, France, Mediterranean Sea	+	– (2)

Table 1. Cont.

Recipient	Life phase, origin	*Eurychasma*	*Chytridium*
Scytosiphonales			
Endarachne binghamiae J. Agardh	Chile	+	+
Petalonia fascia (O. F. Müller) Kuntze	France, Mediterranean Sea	+	– (2)
Scytosiphon lomentaria (Lyngbye) Link	Japan	+	+
	Australia	+	– (2)
Desmarestiales			
Desmarestia distans (C. Agardh) J. Agardh	Patagonia	+ g, sp.	+ g
Desmarestia munda Setchell et Gardner	g, Vancouver Island, Canada	+	– (2)
Desmarestia viridis (O.F. Müller)	g, Newfoundland	+	+
Laminariales			
Laminaria digitata (Hudson) Lamouroux	Brittany	+ g, sp.	– (2)
Lessonia trabeculata Villouta et Santelices	g, Chile	+ g, sp.	+ on oogonia
Macrocystis pyrifera (L.) C. Agardh	g, Chile	+	– (2)
Saccorhiza dermatodea (De la Pylaie) J. Agardh	g, Newfoundland	+	– (2)
Affiliation unsettled			
Microzonia velutina (Harvey) J. Agardh	New Zealand	– (2)	– (2)

+, infection successful; –, no infection of new host; g, gametophyte; sp, sporophyte (given where known). Figures in parentheses give the number of independently repeated assays with negative result.

with aliquots of a unialgal clonal culture of *P. littoralis* (Pil IR) from Isla Diego Ramirez (Chile, Drake Passage, Müller and Stache 1989). Within several weeks the parasites became established on the new host. In a subsequent step, two Pil IR subclones were initiated which were infected by the parasites in the following way: our *Eurychasma* donor culture started with a Pil IR fragment infected with spores from one *E. dicksonii* sporangium, while our *Chytridium* donor culture originated from a Pil IR fragment exposed to spores from 14 *C. polysiphoniae* sporangia. The two host–parasite cocultures were stable and were used for all the infection experiments. They were maintained in plastic Petri dishes filled with Provasoli enriched seawater (PES) medium (Starr and Zeikus 1993) prepared from artificial seawater (WIMEX, Krefeld, Germany, adjusted to 30% salinity), illuminated with daylight-type fluorescent lamps at an irradiance of 9 $\mu E\ m^{-2}\ s^{-1}$ for 10 h day^{-1} and kept at 10 ± 1°C. Cultures were transferred to fresh medium in 1- or 2-week intervals. Recipient host cultures were unialgal clonal isolates of various brown algae originating from the coasts of the countries listed in Table 1. They were routinely maintained in our laboratory as stock cultures and were kept under the same conditions as the *Pylaiella* parasite donor cultures during this study.

Infection experiments were carried out by coincubation of the Pylaiella parasite donor cultures with the prospective host species in 60 mm plastic Petri dishes. Cultures were observed daily under a Leitz (ELVAR) stereo microscope (Wetzlar, Germany) at 63× magnification. A new host was classified as susceptible if sporangia of the pathogen developed on it within 3 weeks. If infection of a new host failed, observation was continued for one more week and the experiment was repeated at least once.

In order to study the temperature dependence of pathogen performance, infection experiments were made at different temperatures with *P. littoralis* and *Acinetospora crinita* (Carmichael) Kornmann as recipients. Plastic Petri dishes containing 10 mL culture medium were placed in different positions along a temperature gradient generated by defined heating and cooling from two ends of an aluminum plate. Experimental temperatures increased in increments of 2.2–2.5°C and were stable within ± 0.5°C of the set value. Infection cycles of the parasites were quantitated as the number of days after which the first empty sporangium was observed on a newly infected algal host.

RESULTS

The cultures confirmed the holocarpic character of both *C. polysiphoniae* and *E. dicksonii*. Infection symptoms remained localized, did not spread to neighboring cells and the parasitic sporangia were formed directly without a mycelium.

The ectoparasite *C. polysiphoniae* occurred on its original host *P. littoralis* (Ectocarpales) at high density, often with two or more sporangia living on one host cell (Fig. 1). Upon maturity, spores escaped through a circular operculum (not shown). In the current infection experiments *Chytridium* attacked the distal uniseriate parts of polystichous thalli, like *Haplospora globosa* Kjellman (Tilopteridales, Fig. 2), and the filamentous microthalli of species with heteromorphic life histories, like the gametophyte of *Carpomitra costata* (Stackhouse) Batters (Sporochnales, Fig. 3). In all cases, *Chytridium* was connected to the host cell by a haustorium. Infected host cells became depleted and eventually died. On all hosts, *Chytridium* sporangia were spherical and reached different sizes depending on the host (Figs 1–3). Generally, *Chytridium* seemed to prefer actively growing uniseriate filaments and was rarely seen on older or senescent cells or on compact thalli.

The endoparasite *E. dicksonii* showed great affinity for its original host, *Pylaiella* (Fig. 4). The cytoplasm of an infected *Pylaiella* cell was completely consumed by the parasite. Its wall expanded and eventually formed an *Eurychasma* sporangium with one or more beak-like extensions which differentiated into apertures for spore release. Host thalli easily fragmented by breakage at the sites of empty parasite sporangia. In *Pylaiella* and many other hosts, empty *Eurychasma* sporangia showed the 'net' structure (Figs 4,6,9,10,12,13), a characteristic ornamentation of the sporangium wall mentioned by earlier workers (Sparrow 1960, p. 818; Jenneborg 1977).

Eurychasma attacked filament cells of the Ectocarpales, the filamentous microthalli of heteromorphic hosts (Figs 5–9) and juvenile sporophytes of *Desmarestia* (Fig. 8). Furthermore, it infected individual cells within compact macrothallus tissue in members of the orders Dictyosiphonales (Fig. 10) and Laminariales (Figs 11, 13,14) and their microscopic gametophyte stages (Figs 12,15). *Eurychasma* also invaded the prominent apical cells as well as subapical and differentiated tissue cells in members of the order Sphacelariales (Figs 16–18).

Table 1 summarizes the results of the infection experiments. A total of 48 host species from 13 orders of the Phaeophyceae were examined. Of these, 45 species were susceptible to infection by *Eurychasma* and 23 to *Chytridium*. Only *Microzonia velutina* (Harvey) J. Agardh was resistant to both parasites. In *Ectocarpus fasciculatus* Harvey, *Hincksia mitchelliae* (Harvey) Silva, *Feldmannia irregularis* (Kuetzing) Hamel and *Scytosiphon lomentaria* (Lyngbye) Link isolates of different geographic origin showed differential susceptibility to the two pathogens.

Figure 19 shows the effect of temperature on the infection of *P. littoralis* and *Acinetospora crinita* by *Eurychasma* and *Chytridium*. Judged by visual inspection of inoculum development, both algae showed maximum growth around 15°C and died at 23°C. *Eurychasma* infected both host species from +4°C to their lethal limit at 23°C. The most rapid development of the parasite occurred at 12°C. At this temperature the complete developmental cycle of *Eurychasma* required 14–15 days in *Pylaiella* and 1–2 days more in *Acinetospora*. *Chytridium* had a narrower temperature range, it persisted on both hosts between 4 and 15°C. At 15°C, with maximum growth rates of the algal hosts, the virulence of *Chytridium* appeared much reduced and only a few fungal sporangia were seen; at 16°C the parasite died. Development of *Chytridium* depended strongly on the host. It was most rapid in *Pylaiella* at 11°C, where a full cycle took 6 days, while the shortest cycles in *Acinetospora* required 11 days at 10°C.

DISCUSSION

The present study represents the first experimental investigation of eukaryotic parasites in brown algae under laboratory conditions. Both *Chytridium* and *Eurychasma* can infect many hosts within a broad taxonomic range and from many coasts of the world. This characterizes them as ubiquitous and cosmopolitan marine pathogens.

In our experiments the severity of infection symptoms differed between various hosts and, occasionally, also in repetitions of the same combination. This may reflect interference of the epibiontic bacterial flora with the settlement of parasite spores and other unknown factors that may modify the virulence of the parasites and the susceptibility of the hosts. Therefore, the negative results in the present study (i.e. resistance of certain hosts to infection) must be considered with caution as

Figs 1–18. Disease symptoms caused by *Chytridium polysiphoniae* and *Eurychasma dicksonii* in laboratory cultures of various brown algal hosts. 1–3. *Chytridium* infection on various hosts 1. Parasite sporangia on *Pylaiella littoralis*. Host cells are attacked by two or more parasites (arrowheads). 2. *Chytridium* on a uniseriate apical part of the *Haplospora globosa* thallus. 3. *Chytridium* on a gametophyte of *Carpomitra costata*. 4–18. *Eurychasma* infection of various hosts. 4. *Pylaiella littoralis*. Host cells with young parasite syncytia (arrowheads) and empty sporangia of the parasite. 5. Microthallus of *Scytosiphon lomentaria*. 6. Microthallus of *Petalonia fascia*. 7. Heavy infestation of older gametophyte culture of *Desmarestia distans* with young sporophytes (arrowheads indicate empty parasite sporangia). 8. A parasite syncytium in a cell of a *D. distans* sporophyte. 9. Parasite sporangium on a gametophyte of *D. distans*. 10. An empty parasite sporangium on *Desmotrichum undulatum*. 11. Host cells of *Lessonia trabeculata* with young parasite syncytia and empty sporangium in juvenile host sporophyte. 12. Parasite sporangia on host gametophyte of *L. trabeculata*. 13. Cells of juvenile sporophyte of *Laminaria digitata* with various stages of parasite development, including fragmentation of the host (14). 15. Parasite in gametophyte cells of *Saccorhiza dermatodea*. 16. The parasite is killing apical and subapical cells of *Stypocaulon scoparia* (arrowheads). Damage by loss of apical cell (asterisk) repaired by regeneration. Immature and empty parasite sporangium emerging from a thallus cell of *S. scoparia* (16, center and 17). 18. An early parasite stage developing in an apical cell of *Halopteris paniculata*. Figs 1,4 from living material, all others from permanent voucher slides of the current infection experiments. Bar = 100 µm applies to Figs 1–3, 4–18, respectively.

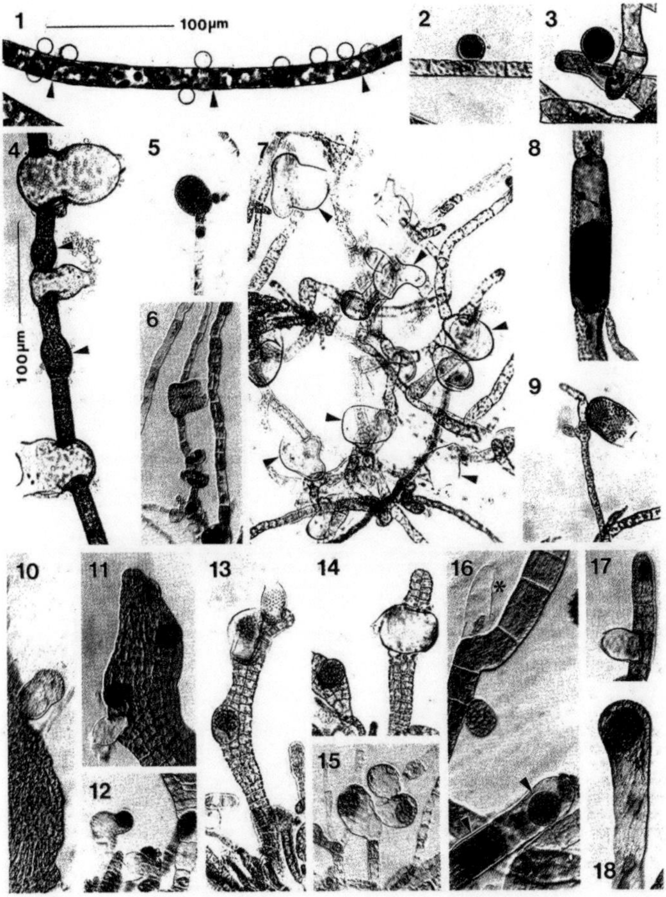
1
100µm
2
3
4
100µm
5
6
7
8
9
10
11
12
13
14
15
16
*
17
18

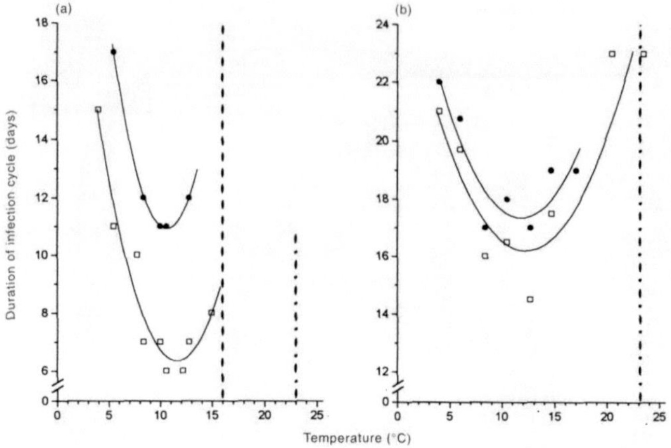

Fig. 19. Correlation between temperature and development of (a) *Chytridium* and (b) *Eurychasma* on the hosts *Pylaiella littoralis* (□) and *Acinetospora crinita* (●). Infection cycle: minimum number of days required for the maturation and discharge of newly formed parasite sporangia. Data were processed by the program Microcal Origin 4.1, and graphs were generated by square regression. Lethal limits for *Chytridium* (----) and algal hosts (------).

the true host ranges of both pathogens may be even broader. Nevertheless, it seems obvious that *Eurychasma* has a significantly broader host range than *Chytridium*.

None of the host cultures in the present infection experiments was killed by *Eurychasma* or *Chytridium* and all host–pathogen combinations arrived at a state of coexistence. We noticed two obvious mechanisms that allowed the hosts to escape from parasite attack. The fragmentation of host thalli caused by *Eurychasma* produced vegetative propagules, which gave the host the chance to drift to new habitats with no or reduced parasite pressure. Mass developments of drifting *Pylaiella* have been attributed to this effect (Wilce *et al.* 1982). The upper lethal limit of 16°C for *Chytridium* allows many hosts to survive and grow without the parasite at higher temperatures. This effect is likely to create seasonal fluctuation of the host–pathogen interaction in natural habitats. We did not observe resting stages or sexuality in the current parasite cultures.

Several brown algal orders exhibit heteromorphic life histories: Chordariales, Dictyosiphonales Scytosiphonales, Scytothamnales, Desmarestiales, Sporochnales and Laminariales. In these groups either the gametophytes or sporophytes are microscopic in size, but constitute obligatory parts of the life cycles. It is difficult or impossible to detect, identify and study such microscopic stages in their natural habitats and, consequently, reports of infections on such stages are completely lacking in the phycological literature. Our study has now revealed that parasitic pressure on the microscopic stages of heteromorphic life histories is likely to occur. It is presently impossible to fully comprehend the consequences of this finding. Depending on circumstances, parasite attack on these cryptic stages may reduce or stimulate the recruitment of macrothalli.

For methodical reasons, the present study could not cover the entire taxonomic range of the Phaeophyceae and members of the Fucales and Durvillaeales with economically and ecologically important representatives were not included. Other groups, like Notheiales or Ascoseirales, were not available for the experiments. Further studies are needed to determine if, and to what degree, these groups are also subject to parasite attack.

ACKNOWLEDGEMENTS

We are grateful to B. Kloareg and S. Loiseaux-de Goër (Roscoff) for discussions and laboratory facilities. F.K. and H.K. were supported by fellowships from

Studienstiftung des Deutschen Volkes (Bonn) and Hüls-AG Stiftung (Marl).

REFERENCES

Dangeard, P. 1934. Sur la présence à Roscoff d'une Chytridiale parasite des Ectocarpacées, l'*Eurychasma dicksonii* (Wright) Magnus. *Ann. Protist.* **4**: 69–73.

Feldmann, J. 1954. Inventaire de la Flore Marine de Roscoff: Algues, champignons, lichens et spermatophytes. *Travaux La Station Biologique Roscoff* **6** (Suppl.): 1–152.

Jenneborg, L. H. 1977. *Eurychasma*-infection of marine algae. Changes in algal morphology and taxonomical consequences. *Bot. Marina* **20**: 499–507.

Karling, J. S. 1943. The life history of *Anisolpidium ectocarpii* gen. nov. et sp. nov., and a synopsis and classification of other fungi with anteriorly uniflagellate zoospores. *Am. J. Bot.* **30**: 637–48.

Karling, J. S. 1977. Chytridiomycetarum *Iconographia J. Cramer.* Vaduz, Monticello, 414 pp.

Kawai, H. and Tokuyama, M. 1995. *Laminarionema elsbetiae* gen. et. sp. nov. (Ectocarpales, Phaeophyceae), a new endophyte in *Laminaria* sporophytes. *Phycol. Res.* **43**: 185–90.

Konno, K. and Tanaka, J. 1988. *Eurychasma dicksonii*, a parasitic fungus on *Acinetospora crinita* (brown alga) in Japan. *Bull. Nat. Sci. Mus. Tokyo Ser. B* **14**: 119–22.

Küpper, F. C. and Müller, D. G. 1999. Massive occurrence of the heterokont and fungal parasites *Anisolpidium, Eurychasma* and *Chytridium* in *Pylaiella littoralis* (Ectocarpales, Phaeophyceae). *Nova Hedwigia*: in press.

Müller, D. G., Kapp, M. and Knippers, R. 1998. Viruses in marine brown algae. *Adv. Virus Res.* **50**: 49–67.

Müller, D. G. and Stache, B. 1989. Life history studies on *Pilayella littoralis* (L.) Kjellman (Phaeophyceae, Ectocarpales) of different geographical origin. *Bot. Mar.* **32**: 71–8.

Peters, A. F. and Ellertsdóttir, E. 1996. New record of the kelp endophyte *Laminarionema elsbetiae* (Phaeophyceae, Ectocarpales) at Helgoland and its life history in culture. *Nova Hedwigia* **62**: 341–9.

Petersen, H. E. 1905. Contributions à la connaissance des phycomycètes marins (Chytridinae Fischer). *Kgl. Danske Videnskabernes Selskabs Forhandlinger* **5**: 439–88.

Siemer, B. L. and Pedersen, P. M. 1995. The taxonomic status of *Pilayella littoralis, P. varia* and *P. macrocarpa* (Pilayellaceae, Fucophyceae). *Phycologia* **34**: 257–66.

Silva, P. C., Lamy, D., Loiseaux-de Goër, S. and de Reviers, B. 1999. Proposal to conserve the name *Pylaiella* Bory (Phaeophyceae) with a conserved spelling. *Taxon* **48**: 1–2.

Sparrow, Jr F. K. 1960. *Aquatic Phycomycetes.* 2nd edn. The University of Michigan Press, Ann Arbor, 1187 pp.

Starr, R. C. and Zeikus, J. A. 1993. UTEX: The culture collection of algae at the University of Texas at Austin. *J. Phycol.* **29** (Suppl.): 1–106.

Wilce, R. T., Schneider, C. W., Quinlan, A. V. and van den Bosch, K. 1982. The life history and morphology of free-living *Pilayella littoralis* (L.) Kjellman (Ectocarpaceae, Ectocarpales) in Nahant Bay, Massachusetts. *Phycologia* **21**: 336–54.

4

Phylogenetic affinities of two eukaryotic pathogens of marine macroalgae, *Eurychasma dicksonii* and *Chytridium polysiphoniae*

Frithjof C. Küpper[1), 2) #)], Ingo Maier[2)], Dieter G. Müller[2)], Laure Guillou[3)], and Susan Loiseaux-de Goër[1)]

[1)] UMR 1931 CNRS-Goëmar, Station Biologique, Place G. Teissier, BP 74, F-29682 Roscoff, Bretagne, France

[2)] Dept. of Biology, Faculty of Sciences, University of Konstanz, D-78457 Konstanz, Germany

[3)] Institut de Ciències del Mar, Passeig Joan de Borbó, s/n- 08039 Barcelona, Spain.

[#)] To whom correspondence should be addressed. E-mail: kuepper@sb-roscoff.fr
Fax: +33 – 2 98 29 23 24   Phone: + 33- 2 98 29 23 62

Running title

Molecular phylogeny of *Eurychasma* and *Chytridium*

Key words

chitin - *Chytridium* – Chytridiomycota - *Eurychasma* – Oomycota – *Pylaiella* – SSU rDNA

Abstract

The 18S rRNA genes of *Eurychasma dicksonii* and *Chytridium polysiphoniae*, pathogens of brown algae, were sequenced and used to clarify their phylogenetic affiliations. *E. dicksonii* appears isolated at the basis of the Oomycota, intermediate between terrestrial and freshwater organisms on the one side, and free-living marine plankton organisms on the other. In contrast, *C. polysiphoniae* clearly aligns with the Eumycota, in either the *Rhizophydium* or the *Spizellomyces* clade of the Chytridiomycota, but clearly not with other *Chytridium* species for which sequences are available. In coherence with these findings, *C. polysiphoniae* has a chitin cell wall, whilst *E. dicksonii* has cellulose instead.

Introduction

Eurychasma dicksonii (Wright) Magnus and *Chytridium polysiphoniae* Cohn are two widespread pathogens of marine algae (Sparrow, 1960). On European coasts, they occur as massive epidemics in *Pylaiella littoralis* (L.) Kjellman (Küpper and Müller, 1999 / Chapter 2). Field observations indicated that they have a wide host range (Sparrow, 1960; Jenneborg, 1977), which was recently confirmed in laboratory cultures (Müller et al., 1999 / Chapter 3).

Both pathogens are holocarpic, i.e. a single host cell is infected by a spore. Without formation of hyphae and infection of neighboring cells, the infection directly results in a sporangium.

From their very first discovery, there has been considerable confusion about the phylogenetic relationships of *Eurychasma* and *Chytridium*, and they have repeatedly been placed together into the same groups. Wright (1879), in the first description of the species *Eurychasma dicksonii* (Wright) Magnus, names it *Rhizophydium dicksonii*, but speaks of it as a "*Chytridium*", citing Braun (1844): "The Chytridia form a new genus of unicellular, parasitical Algae, or, if it be preferred, of aquatic Fungi". "Chytridia" in this original sense were actually a polyphyletic group of all aquatic plant pathogens known at that time.

Rattray (1885) called *Rhizophydium dicksonii* Wright a "chytridiaceous parasite". Shortly later, Hauck (1878) wrote "diese neue Chytridiacee", whilst Wille (1899) tried to include it into the genus *Olpidium* as *O. dicksonii* (Wright) Wille, but this was not widely accepted. Magnus elevated *Rhizophydium dicksonii* to the status of a genus of its own with the name *Eurychasma*.

Sparrow (1934) did not yet draw a clear separation between fungi (Eumycota) and Oomycota (Heterokonta) among aquatic plant pathogens and used the terms "Phycomycetes" and "Chytridineae". Nevertheless, he commented, "whether we regard *Eurychasma* as a filamentous phycomycete which has been greatly reduced and modified by its mode of life, or whether we look upon it as a primitive form foreshadowing tendencies found in the higher groups, in the structure and behavior of its zoospore and its possession of cellulose walls, it exhibits distinct affinities with the Saprolegniales and not with the Chytridiales to which it is usually referred." All subsequent authors (e.g. Aleem, 1950; Feldmann, 1954; Sparrow, 1960; Konno and Tanaka, 1988) placed *Eurychasma* within the Saprolegniales.

Chytridium polysiphoniae was first described by Cohn (1865) on *Polysiphonia violacea* in Helgoland. Petersen (1905) changed it to *Rhizophydium polysiphoniae* (Cohn) Petersen, but Sparrow (1934) conserved the original nomenclature. Feldmann (1954) and Sparrow (1960) placed this pathogen in the Chytridiaceae, Chytridiales.

Rhizophydium is nowadays classified among the Eumycota (e.g. James et al., 2000), whilst *Eurychasma* has been placed in the Eurychasmataceae, Myzocytiopsidales, Oomycota, Heterokonta (Hawksworth et al., 1996). The same authors classify the genus *Chytridium* in the Chytridiaceae, Chytridiomycota (Eumycota).

In this study, the small subunit ribosomal RNA genes (18 S rDNA) were sequenced and used to examine the phylogenetic affiliations of these two pathogens. To our knowledge, no molecular work has been done so far on any marine oomycete. Furthermore, a recently developed technique (Maier et al., 2000) was used to address the occurrence of chitin in the cell walls of these pathogens, und UV fluorescence microscopy to detect autofluorescent compounds in infected algal cells and the pathogens themselves.

Materials and Methods

Biological material

P. littoralis infected by both *C. polysiphoniae* and *E. dicksonii* was collected at Aith Voe (Bressay, Shetland) in April 1996 (Küpper and Müller, 1999 / Chapter 2). Unialgal host cultures were established as described previously (Müller et al., 1999 / Chapter 3). Briefly, small tufts of infected *Pylaiella* were co-incubated with aliquots of a unialgal clonal culture of *P. littoralis* (Pyl IR) from Isla Diego Ramirez (Chile, Drake Passage; Müller and Stache, 1989). Within several weeks, *C. polysiphoniae* and *E. dicksonii* became established on the new host. In a subsequent step, two Pyl IR subclones were initiated, each infected by one of the parasites only in the following way: 1) culture Eu Pyl IR 1 starting from spores of one *E. dicksonii* sporangium and 2) culture Chyt Pyl IR 14 starting from spores of 14 sporangia of *C. polysiphoniae*. These unialgal associations of the host *P. littoralis* and either *E. dicksonii* or *C. polysiphonae* as parasites were used for all our experiments.

Cultures were maintained in plastic Petri dishes in Provasoli ES (Starr & Zeikus, 1993) prepared from filtered natural seawater (collected off Roscoff, Brittany). They were illuminated with daylight-type fluorescent lamps at an irradiance of 9 $\mu E\ m^{-2}\ s^{-1}$ for 10h day^{-1} and kept at 10 ±1°C. All cultures were transferred to fresh medium in one or two week intervals.

Chitin cytochemistry

Overall, the method described recently by Maier et al. (2000) was applied. Parasitized algal filaments were fixed in ethanol/acetic acid (3 : 1). After two short washing steps in 70% ethanol, the material was equilibrated in phosphate-buffered saline (PBS, 13.7 mM NaCl, 3 mM KCl, 8.1 mM Na_2HPO_4 x 2 H_2O, 1,5 mM KH_2PO_4, pH 7.5) for 10 min. Unspecific protein binding sites were blocked by incubation on a shaker with 3 % bovine serum albumin (BSA, Sigma) in PBS containing 0.05 % sodium azide for 1 hour at room temperature. After removal of the blocking solution, a fluorescein isothiocyanate (FITC) - conjugated recombinant chitin binding protein (FUNGALASE™-F, Anomeric, Baton Rouge, USA) was applied according to the manufacturer´s protocol (1:10 dilution), but omitting the periodic acid oxidation step. The staining reaction was carried out for 2 h at room temperature in the dark, followed by two washes with PBS for 5 min each. Afterwards, the specimens were mounted in SLOWFADE-LIGHT antifade solution (MoBiTec, Göttingen, Germany). Photographs were taken on Ilford 400 ASA HP5 Plus film (fluorescence) and Kodak 50 ASA Technical Pan film (bright field), using blue excitation light.

Cloning and sequencing of SSU rRNA genes

Infected algal material (either Eu Pil IR 1 or Chyt Pyl IR 14) was dried with silicagel. Micro-extractions were carried out by grinding 1-2 mg dry weight with a micro-pestle in an Eppendorf tube, to which a tiny spatule (< 50 mg) of Fontainebleau sand, 600 µl of extraction buffer (25% sucrose, 50 mM TRIS, 1 mM EDTA, pH 7.5) and repeatedly small quantities of liquid nitrogen were added. Grinding was carried out until the whole content of the tube had a homogenous, yellow-brownish appearance. 10 µl (6 U) of proteinase K (Boehringer Mannheim, Germany) were added and the mixture was incubated at 37°C for two hours. After addition of 1 volume of phenol, the tube was gently shaken for 10 min. The aqueous phase containing the DNA was recovered after centrifugation at 12500 g for 15 min. Again, an isovolume of chloroform and phenol was added and gently mixed to obtain a single phase. The aqueous phase was recovered after centrifugation and the DNA was precipitated by addition of 1/10 volume of 3 M sodium acetate (pH 7.5) and 2 volumes of cold (- 20 °C) ethanol (analytical grade). After 30 min at –80°C and centrifugation at 12500 g for 25 min, the supernatant was removed and the pellet was washed by addition of 70 % ethanol and centrifugation. After air-drying, the pellet was re-suspended in 500 µl H_2O and further purified using PhytoPure™ resin (Nucleon, Amersham Life Sciences, Little Chalfont, Buckinghamshire, England). The purified DNA was re-suspended in 50 µl sterile water.

PCR was carried out using 1 µl template, 5 µl (200 pmol/µl) of each oligonucleotide primer # 328 (5'-ACCTGGTTGATCCTGCCAG-3') and # 329 (5'-TGATCCTTCYGCAGGTTCAC-3'), respectively, 14 µl sterile water (i.e. 25 µl in total) and one Ready-to-go PCR bead™ (Amersham-Pharmacia) per reaction. The primers were complementary to conserved sequences close to the respective 5' and 3' termini of the 18S rRNA gene, designed to amplify most eukaryotic 18 S rRNA genes (Moon – van der Staay *et al*., 2000). Dilution (1/10 or 1/100) of the template proved to have a beneficial effect on PCR efficiency, presumably due to dilution of residual carbohydrate and polyphenol contaminants originating from the brown algal tissue. After verification of purity of the PCR product, cloning was carried out immediately using the TOPO TA Cloning® vector (Invitrogen®) in *Escherichia coli*. Plasmid DNA was prepared using the Flexiprep™ kit (Pharmacia). Due to the mixed extraction of both *Pylaiella* and *Eurychasma* or *Chytridium* DNA, respectively, the clones obtained were inevitably a mixture of brown algal and pathogen 18 S rDNA. In each case, the two kinds of clones were distinguished by their different restriction patterns obtained by a combined Eco RI and Bam HI digestion. Sequencing (double strand) was carried out by ESGS-Cybergene (Evry, France), using a primer-walking approach and yielding three different consensus sequences.

Phylogenetic trees

The three obtained sequences were manually aligned with other heterokont and fungal taxa, respectively, taking secondary structures into account. Bases which occurred in hypervariable regions and which could not be aligned with confidence were removed yielding an alignment of 1746 positions for the heterokonts and 1034 for the fungi (the alignments are available from the corresponding author, F.C.K.). These alignments were used to generate distance and parsimony phylogenetic trees. Distance analysis (Neighbor-Joining) were performed using the PHYLIP™ package (release 3.57c; Felsenstein, 1993). The Kimura two-parameter option was employed to compute evolutionary distances (Kimura 1980). Bootstrap analyses (100 replicates, Felsenstein, 1985) were performed for both cases. Several different trees including more or less species or taking different outgroups (*Opalina ranarum*, belonging to the Opalinids (AF141970) for the heterokonta and *Allomyces macrogynus*, belonging to the Blastocladiales (U23936) for the fungi) were done in each case without changing the results.

Results and Discussion

Our light-microscopic observations of *E. dicksonii* and *C. polysiphoniae* corresponded well to those made earlier by a number of authors (Wright, 1879; Rattray, 1885; Aleem, 1950; Jenneborg, 1977 for *Eurychasma*; Petersen, 1905, for *Chytridium*; Sparrow, 1960, and Küpper and Müller, 1999 / Chapter 2, for both).

Histochemistry

Whilst the vast majority of fungi have chitin in their cell walls, its occurrence in the heterokont Oomycota is not uniform (Barr, 1983). Unlike some species such as *Achlya radiosa* (Campos-Takaki et al., 1982), *Eurychasma* does not produce chitin, as demonstrated by the absence of fluorescence upon Fungalase™ staining. Sparrow (1934) did not mention the absence of chitin in *Eurychasma*, but he already based his argument to place the species within the Saprolegniales upon his findings of cellulose in its cell wall (determined by application of zinc chloriodide). In contrast, we could demonstrate that the fungus *Chytridium* does have chitin in its cell wall (Fig. 1).

In the first half of the infection cycle, *Eurychasma*-infected *Pylaiella* cells did not show any accumulation of UV-autofluorescent compounds and chlorophyll fluorescence remained seemingly unchanged (Fig. 2 A), indicating that the impact of the oomycete is rather modest at first and that photosynthesis is unimpaired. The host cell is killed towards the end of the infection cycle, accompanied by strong blue autofluorescence (Figs 2 B, C). In contrast, UV fluorescence microscopy revealed the presence of autofluorescent compounds in the cell wall of unstained *Chytridium* (Figs 2 D-E). Infected *Pylaiella* cells also accumulated fluorescent compounds – possibly polyphenolics –, probably located in physodes (which often accumulate in old or the senescent cells). Following infection, *Pylaiella* cells usually died rapidly as revealed by the loss of chlorophyll autofluorescence (Fig. 2 F) in heavily infested branches (Fig. 2 G). Sparrow (1934), having found *C. polysiphoniae* in the red alga *Ceramium fruticulosum*, already observed that this fungus was living on dead or dying cells, but, contrasting our views of *C. polysiphoniae* as an obligate biotrophic fungus, attributed this rather to saprophytism rather than to the fungus killing its host. Results concerning the impact of *Eurychasma* and *Chytridium* on photosynthetic performance of their host at the cellular level throughout their infection cycles are discussed elsewhere (Chapter 5).

Nucleotide sequences

In total, 3 different 18 S rDNA sequences were obtained from the infected cultures: each culture (Eu Pyl IR 1 and Chyt Pyl IR 14, respectively) yielded two different clones. One clone each from each culture had the same Eco RI and Bam HI restriction pattern, while the two other ones were distinctly different (not shown).

Manual alignments showed that one sequence possessed numerous signatures characteristic of Oomycota, and could be attributed to *Eurychasma.* Its restriction pattern was found only in one of the two types of clones from Eu Pyl IR 1, but not in those from Chyt Pyl IR 14.

The second sequence in the pool of 18S rDNA clones could be attributed to *Pylaiella*, having clear brown algal signatures. According to its restriction pattern, it was present in the clones from both Eu Pyl IR 1 and Chyt Pyl IR 14.

In contrast, the third sequence clearly aligned with the Eumycota, and was attributed to *Chytridium*. Its restriction pattern was found only in one of the two types of clones from Chyt Pyl IR 14, but not in those from Eu Pyl IR 1.

Using the alignments and secondary structure to determine the 5' and 3' ends, the 18S rDNA sequences of *Pylaiella littoralis*, *Eurychasma dicksonii* and *Chytridium polysiphoniae* were found to be 1823, 1820 and 1808 bp long, respectively. Their partial sequences (excluding primers) have been deposited in GenBank and assigned the accession numbers AY032606 (*P. littoralis*, 1780 bp), AY032607 (*E. dicksonii*, 1777 bp) and AY032608 (*C. polysiphoniae*, 1766 bp). Herbarium specimens (microscope slides) have been deposited in the Jepson Herbarium (University of California, Berkeley / UC; *E. dicksonii*: UC 1726827, *C. polysiphoniae*: UC 1726828), the National Herbarium of Victoria (Melbourne / MEL; *E. dicksonii*: MEL 2068385, *C. polysiphoniae*: MEL 2068352), in the Muséum National d'Histoire Naturelle – Cryptogamie (Paris, PC – this collection also received dried specimens in silicagel), and the collection of CAB International (Egham, Surrey, U.K.; *E. dicksonii*: IMI 385979, *C. polysiphoniae*: IMI 385980).

Manual alignments already revealed that *Eurychasma* had typical signatures of oomycetes as well as of more archaic heterokonts (Fig. 3). The Neighbor-Joining (N-J) tree shown in Fig. 4 contains 45 species of heterokonts including 12 oomycetes. In this tree and in all N-J trees obtained which contained more species of heterokont algae (Chrysophyceae and Synurophyceae, Eustigmatophyceae, Raphidophyceae, Pedinellophyceae, Dictyochophyceae, Pelagophyceae) and more heterotrophic heterokonts (Proteromonas, Thraustochytrids, more Labirinthulids and Opalinids as a root), the oomycete lineage splits in two sister branches, consistent with observations by Petersen and Rosendahl (2000). *Eurychasma* always branches from this lineage before its separation into two branches, with a bootstrap value of 100 %. Parsimony trees showed the same topology for the oomycete lineage except for the position of *Apodachlya brachynema* which branched with the *Leptolegnia* clade and not with the *Lagenidium* clade. Therefore *Eurychasma* appears to have differentiated early, before the radiation of other known oomycetes which are grouped in two lineages as suggested originally by Dick et al. (1984) and confirmed by Petersen and Rosendahl (2000). Interestingly, according to the distance tree, *Eurychasma* is about equally distant and intermediate from the well-known terrestrial and fresh water oomycetes on the one side, and two members of the eukaryotic, oceanic picoplankton (OLI 11008, OLI 11026) on the other side which, in our own N-J trees and those of Moon – van der Staay et al. (2001), branch at the base of the oomycete lineage. It is not known if these are free-living or parasitic organisms (Moon – van der Staay et al., 2001). Parsimony yielded a slightly different result, placing these oceanic sequences on the same branch as *Developayella elegans*, diverging just before the oomycete lineage, without changing the position of *Eurychasma*. Sparrow's (1934, 1960) classification of *Eurychasma* in the order Saprolegniales can clearly not be maintained.

Our findings add an interesting aspect to the evolution of the Oomycota: Barr (1983) had suggested that terrestrial plant pathogens had evolved from aquatic saprobes belonging to the Saprolegniales. Our results show that the Oomycota comprise early emerged pathogens of marine algae other than Chlorophyta, suggesting that obligate parasitism has evolved earlier than considered so far in the lineage of the Oomycota.

James et al. (2000) recently carried out a comprehensive molecular study on the systematics of the Chytridiomycota based on ribosomal RNA genes. Our analysis, based on parsimony and Neighbor-Joining, confirmed their repartition into different clades (with a clade each of *Lacustromyces*-, *Rhizophydium*-, *Spizellomyces*-, *Nowakowskiella*-, *Monoblepharis*- and *Chytridium*-like organisms) and their possibly polyphyletic origin. The sequence analysis showed that *Chytridium polysiphoniae* is not closely related to the two other members of the genus *Chytridium* for which sequences are available, *Chytridium confervae* and *Chytridium* sp. strain DU-DC2 or to the clade which comprises these. This result confirms the statement of James et al. (2000) that the Chytridiales, in their current classification, may not be monophyletic. We conclude that the position of *C. polysiphoniae* in this genus is unjustified – contrary to Sparrow (1934), arguing that "this fungus because of the operculate character of the sporangium would seem to be better placed in the genus *Chytridium* than in *Rhizophydium*". Sparrow (1934) himself obviously wanted to correct Petersen (1905), who had renamed the species *Rhizophydium polysiphoniae* (Cohn) Petersen. Our results indeed seem to confirm Petersen (1905). Alignments indeed do reveal characteristic, *Rhizophydium*-like, but not *Chytridium*-like signatures (Fig. 5). Also the consensus of parsimony (Fig. 6) places it in the *Rhizophydium* clade, yet founded on rather weak bootstrap values of 51.3 % (clades according to James et al., 2000). In contrast, Neighbor-Joining places the taxon in the adjacent Spizellomycetales clade (not shown). Based on these results, a re-classification and nomenclatural change of the taxon *Chytridium polysiphoniae* (COHN) PETERSEN appears inevitable. As Barr (1990) and James et al. (2000) suggest in general for the Chytridiomycota, ultrastructural work on zoospores can be expected to contribute to an unambiguous assignment of this aquatic fungus. Therefore, we propose to await further ultrastructural evidence for a final judgement about the taxonomic status of *Chytridium polysiphoniae*.

Conclusions

This study shows that two marine pathogens, *Eurychasma dicksonii* and *Chytridium polysiphoniae*, with wide host ranges among marine macroalgae belong to branches of the Oomycota and Chytridiomycota, respectively. The oomycete *Eurychasma* – the first marine oomycete analysed with molecular methods – branches early from the lineage, before the separation of all known terrestrial and freshwater oomycetes into two clades.

In contrast, *Chytridium polysiphoniae* can be assigned either to the *Rhizophydium* or *Spizellomycetales* clades, but not to the *Chytridium* clade. Ultrastructural studies and a nomenclatural change of this taxon appear necessary.

Our molecular results are confirmed by a histochemical character, which could help to characterize these species from field collections: *C. polysiphoniae* – a fungus – contains chitin, whilst chitin is absent in the heterokont *E. dicksonii*.

Acknowledgements

We would like to thank Catherine Leblanc (Roscoff) for support in molecular work. We gratefully acknowledge Daniel Vaulot (Roscoff) for sharing his sequence data and discussions. FCK was supported by fellowships from Studienstiftung des Deutschen Volkes (Bonn) and the European Commission (program MAST-III).

References

Aleem, A.A. (1950) A fungus in *Ectocarpus granulosus* C. Agardh near Plymouth. Nature 165, 119-120

Barr, D.J.S. (1983) The zoosporic groupings of plant pathogens. Entity of non-entity? In Zoosporic Plant Pathogens. A modern perspective (S.T. Buczacki, ed.): 43-83. London: Academic Press

Barr, D.J.S. (1990) Phylum Chytridiomycota. In Handbook of Protoctista. (L. Margulis, J.O. Corliss, M. Melkonian, and D.J. Chapman, eds): 454-466. Boston: Jones & Bartlett

Braun, A. (1844) Beobachtungen ueber die Verjüngung in der Natur. Freiburg – transl. by Henfrey, A. (1854) The phenomenon of rejuvenescence in nature, especially in the life and development of plants. In: Botanical and Physiological Memoirs, The Ray Society

Campos-Takaki, G.M., Dietrich, S.M., Mascarenhas, Y. (1982) Isolation and characterization of chitin from the cell walls of *Achlya radiosa*. Journal of General Microbiology 128, 207-209

Cohn, F. (1865): Chytridii species novae marinae. Hedwigia 4, 169-170

Dick, M.W., Wong, P.T.W., and Clark, G. (1984) The identity of the oomycete causing "Kikuyu Yellows", with a reclassification of the downy mildews. Botanical Journal of the Linnean Society 89, 171-197

Feldmann, J. (1954) Inventaire de la Flore Marine de Roscoff: Algues, champignons, lichens et spermatophytes. Trav. Stat. Biol. Roscoff, Suppl. 6, Roscoff.

Felsenstein, J. (1993) PHYLIP (Phylogeny Interference Package). Version 3.5c. Department of Genetics, University of Washington

Felsenstein, J. (1985) Confidence limits on phylogenies: an approach using the bootstrap. Evolution 39, 783-791

Hauck, F. (1878) Notiz über *Rhizophydium dicksonii* Wright. Oesterreichische Botanische Zeitschrift 28, 321

Hawksworth, D.L., Kirk, P.M., Sutton, B.C., and D.N. Pegler (1996) Ainsworth & Bisby's dictionary of the fungi. - 8th ed., New York, Oxon: CAB International

James, T.Y., Porter, D., Leander, C.A., Vilgalys, R., and Longcore, J.E. (2000) Molecular phylogenetics of the Chytridiomycota supports the utility of ultrastructural data in chytrid systematics. Canadian Journal of Botany 78, 336-350

Jenneborg, L.-H. (1977) *Eurychasma*-infection of marine algae. Changes in algal morphology and taxonomical consequences. Botanica Marina 20, 499-507

Kimura, M. (1980) A simple method for estimating evolutionary rates of base substitutions through comparative studies of nucleotide sequences. Journal of Molecular Evolution 16, 111-120

Küpper, F.C., and Müller, D.G. (1999) Massive occurrence of the heterokont and fungal parasites *Anisolpidium*, *Eurychasma* and *Chytridium* in *Pylaiella littoralis* (Ectocarpales, Phaeophyceae). Nova Hedwigia 69, 381-389

Magnus P (1905) Über die Gattung zu der *Rhizophydium dicksonii* Wright gehört. Hedwigia 44, 347-349

Maier, I., Parodi, E., Westermeier, R., and Müller, D.G. (2000) *Maullinia ectocarpii* gen. et sp. nov. (Plasmodiophorea), an intracellular parasite in *Ectocarpus siliculosus* (Ectocarpales, Phaeophyceae) and other filamentous brown algae. Protist 151, 225-38

Moon-van der Staay, S.Y., De Wachter, R., and Vaulot, D. (2001) Oceanic 18S rDNA sequences from picoplankton reveal unsuspected eukaryotic diversity. Nature 409, 607-610

Moon-van der Staay, S.Y., van der Staay, G.W.M., Guillou, L., Vaulot, D., Claustre, H., and Medlin, L.K. (2000) Abundance and diversity of prymnesiophytes in the picoplankton community from the equatorial Pacific Ocean inferred from 18S rDNA sequences. Limnology & Oceanography 45(1), 98-109

Müller, D.G., Küpper, F.C., and Küpper, H. (1999) Infection experiments reveal broad host ranges of *Eurychasma dicksonii* (Oomycota) and *Chytridium polysiphoniae* (Chytridiomycota), two eukaryotic parasites in marine brown algae (Phaeophyceae). Phycological Research 47, 217-23

Müller, D.G., and Stache, B. (1989) Life history studies on *Pilayella littoralis* (L.) Kjellman (Phaeophyceae, Ectocarpales) of different geographical origin. Botanica Marina 32, 71-78

Petersen, A.B., and Rosendahl, S. (2000) Phylogeny of the Peronosporomycetes (Oomycota) based on partial sequences of the large ribosomal subunit (LSU rDNA). Mycological Research 104 (11), 1295-1303

Petersen, H.E. (1905) Contributions à la connaissance des phycomycètes marins (Chytridinae Fischer). Kgl. Danske Videnskabernes Selskabs Forhandlinger 5, 439-488

Rattray, J. (1885) Note on *Ectocarpus*. Transactions of the Royal Society of Edinburgh 32, 589-602

Sparrow, F.K. Jr. (1934) Observations on marine phycomycetes collected in Denmark. Dansk Botanisk Arkiv 8(6), 1-24

Sparrow, F.K. Jr. (1960) Aquatic Phycomycetes. 2nd ed. Ann Arbor: The University of Michigan Press, 1187 pp

Starr, R.C., and Zeikus, J.A. (1987) UTEX - the culture collection of algae at the University of Texas at Austin. Journal of Phycology 23 (Suppl.): 1-47

Wille N (1899) Om nogle Vandsoppe. Vidensk. Selskabets Skrifter, Kristiania. Math.-naturw. Kl. 1899, 1

Wright, E.P. (1879) On a species of *Rhizophydium* parasitic on species of *Ectocarpus*, with notes on the fructification of the *Ectocarpi*. Transactions of the Royal Irish Academy 26, 369-379

Tab. 1. GENBANK accession numbers of *Pylaiella littoralis*, *Eurychasma dicksonii* and the other heterokonts used for the alignment (Fig. 2).

Species	**Lineage**	**Abbreviation used in Fig. 3**	**Sequence accession no.**
Synura spinosa	Chrysophyceae		M87336
Ochromonas danica	Chrysophyceae		M32704
Heterosigma akashiwo	Raphidophyceae		L42529
Pelagomonas calceolata	Pelagophyceae		U14389
Pelagococcus subviridis	Pelagophyceae		U14386
Giraudyopsis stellifera	Chrysomerophyceae		U78034
Laminaria japonica	Phaeophyceae		AF123575
Costaria costata	Phaeophyceae		M97958
Desmarestia ligulata	Phaeophyceae		L43060
Scytosiphon lomentaria	Phaeophyceae		L43066
Ectocarpus siliculosus	Phaeophyceae		L17015
Pylaiella littoralis	Phaeophyceae		AY032606
Sargassum macrocarpum	Phaeophyceae		AB011432
Fucus distichus	Phaeophyceae		AB011423
Botrydium stoloniferum	Xanthophyceae		U41648
Tribonema aequale	Xanthophyceae		M55286
Botrydiopsis intercedens	Xanthophyceae		U41647
Bacillaria paxillifer	Bacillariophyceae		M87325
Thalassionema nitzschioides	Bacillariophyceae		X77702
Coscinodiscus radiatus	Bacillariophyceae		X77705
Rhizosolenia setigera	Bacillariophyceae		M87329
Bolidomonas pacifica	Bolidophyceae		AF123595
Bolidomonas mediterranea	Bolidophyceae		AF123596
Oli11008	?	OLI11008	AJ402350
Oli11026	?	OLI11026	AJ402339
Eurychasma dicksonii	Oomycota / this study	E. dicks.	AY032607

Tab. 1 cont.

Apodachlya brachynema	Oomycota	A. brach.	AJ238663
Lagenidium giganteum	Oomycota	L. gigan.	M54939
Phytophthora undulata	Oomycota	P. undul.	AJ238654
Pythium monospermum	Oomycota	P. monos.	AJ238653
Pythiopsis cymosa	Oomycota	P. cymosa	AJ238657
Aplanopsis terrestris	Oomycota	A. terr.	AJ238658
Achlya bisexualis	Oomycota	A. bisex.	M32705
Leptolegnia chapmanii	Oomycota	L. chap.	AJ238661
Leptolegnia chapmanii	Oomycota	L. chap.1	AJ238660
Leptolegnia caudata	Oomycota	L. cau.	AJ238659
Achlya apiculata	Oomycota	A. apic.	AJ238656
Saprolegnia ferax	Oomycota	S. ferax	AJ238655
Developayella elegans	(stramenopiles)	D. eleg.	U37107
Hyphochytrium catenoides	Hyphochytriomycota	H. caten.	X80344
Cafeteria roenbergensis	Bicosoecida		L27633
Siluania monomastiga	Bicosoecida		AF072883
Labyrinthuloides minuta	Labyrinthulida		L27634
Schizochytrium minutum	Labyrinthulida, Thraustochytriidae		AB022108
Thraustochytrium multirudimentale	Labyrinthulida, Thraustochytriidae		AB022111

Tab. 2. GENBANK accession numbers of *Chytridium polysiphoniae* and the other fungi used for the alignment and calculation of trees (Figs 5,6).

Species	**Strain identification**	**Order**	**Abbreviation used in Fig. 5**	**Sequence accession no.**
Karlingiomyces sp.	JEL93	Chytridiales	K. spec.	AF164278-80
207 multiple rhizoids	JEL 207	Chytridiales	207 m. rh.	AF164261-2
Lacustromyces hiemalis	JEL 31	Chytridiales	L. hiemal.	AF164274-5
Polychytrium aggregatum	JEL 190	Chytridiales	P. aggreg.	AF164276-7
Rhizophydium chaetiferum	JEL 39	Chytridiales	R. chaet.	AF164263
Rhizophydium sp.	UGA-F16	Chytridiales	R. sp. F16	AF164264-5
Rhizophlyctis harderi	JEL 171	Chytridiales	R. harder.	AF164272-3
Rhizophydium sphaerotheca	JEL 08	Chytridiales	R. sphaer.	AF164259-60
Rhizophydium sp.	JEL138	Chytridiales	R. sp.138	AF164266-7
Chytridium polysiphoniae	Chyt Pyl IR-gl4	this study	C. poly.	AY032608
Rhizophlyctis rosea	BK47-07	Spizellomycetales	R. ros.47	AF164251-2
Rhizophlyctis rosea	BK57-5	Spizellomycetales	R. ros.57	AF164249-50
Gaertneriomyces semiglobiferus	BK91-10	Spizellomycetales	G. semig.	AF164247-8
Spizellomyces acuminatus	GENBANK	Spizellomycetales	S. acum.	M59759
Spizellomyces kniepii	UGA-F22	Spizellomycetales	S.kniep.	AF164237-8
Powellomyces variabilis	BK91-11	Spizellomycetales	P. var.91	AF164241-2
Powellomyces hirtus	UGA-F18	Spizellomycetales	P. hirtus	AF164239-40
Powellomyces variabilis al.	BK85-1	Spizellomycetales	P. var.85	AF164243-4
Powellomyces sp.	JEL95	Spizellomycetales	P. sp.95	AF164245-6

Tab. 2 cont.

Cladochytrium replicatum	JEL38	Chytridiales		AF164297-8
Allochytridium expandens	BK 69-3	Chytridiales		AF164291-2
Diplochytridium lagenarium	JEL 72	Chytridiales		AF164285-6
Nowakowskiella elegans	BK50-1	Chytridiales		AF164281-1
Harpochytrium sp.	JEL94	Monoblepharidales		AF164331-2
Monoblepharis insignis	BK 59-7	Monoblepharidales		AF164333
Asterophlyctis sarcoptoides	JEL 186	Chytridiales	A. sarc.	AF164317-8
Chytriomyces spinosus	JEL 59	Chytridiales	C. spin.	AF164323-4
Obelidium mucronatum	JEL 57	Chytridiales	O. mucr.	AF164309-10
Phlyctorhiza endogena HANSON	JEL 80	Chytridiales	P. endog.	AF164313-4
Rhizophydium sp.	UGA-F15	Chytridiales	R. sp.15	AF164319-20
Chytridium sp.	DU-DC2	Chytridiales	C. sp.DU	AF164321-2
Chytridium confervae (WILLE) MINDEN	BK M62706	Chytridiales	C. conv.	M59758
Rhizophydium sp.	JEL151	Chytridiales		AF164270-1
Allomyces macrogynus (Emerson) Emerson & Wilson	GENBANK	Blastocladiales		U23936

Figures

Fig. 1. Light microscopy of *Chytridium*-infected *Pylaiella*, treated with Fungalase™ demonstrating the presence of chitin in the cell wall of this fungus: A mature *Chytridium* sporangium during detachment of the operculum (A, B).- Older, empty sporangia – three remaining spores have germinated inside the parental sporangium (C, D). Spores produces chitin soon after settling (the spores themselves are unwalled, e.g. James et al., 2000), causing an intense fluorescence upon Fungalase™ staining and blue excitation (E, F).- In contrast, *Eurychasma dicksonii* does not contain any chitin, causing no fluorescence in Fungalase™-treated, fixed specimens (not shown).- All scale bars correspond to 25 μm.

Fig. 2. In the first half of the infection cycle, *Eurychasma*-infected *Pylaiella* cells do not show any increased accumulation of UV-autofluorescent compounds, indicating that the impact of the intracellular presence of the oomycete is rather modest at first (A), only overwhelming its host towards the end of its infections cycle, accompanied by strong blue fluorescence (B, C).- In comparison, *Chytridium* (D) itself shows blue autofluorescence under UV excitation (E), whilst concomitantly, infected *Pylaiella* cells accumulate increased quantities of fluorescent compounds – possibly polyphenolics – , probably located in the physodes. Following infection, *Pylaiella* cells usually die rapidly as revealed by the loss of chlorophyll autofluorescence (F) in heavily infested branches (G).- All scale bars correspond to 50 μm.

Fig. 3. Partial alignment of heterotrophic heterokonts SSU rDNA sequences, including that of *Eurychasma dicksonii* showing lineage signatures. Eurychasma shows both typical signatures of oomycetes as well as rather archaic signatures of the heterokonts. Numbers indicate the the nucleotides between those which are shown. The included species can be found in table 1.

Fig. 4. Neighbor-Joining tree obtained from alignments (1746 positions) of the SSU rDNA sequences of 45 heterokonts, including *Eurychasma dicksonii*, Oomycota, Bolidophyceae, Bacillariophyceae, Phaeophyceae, and a Chrysophyceae. Bold figures above the branching are the bootstrap values (100 replicates) for Neighbor-Joining, figures in italics below are bootstrap values for parsimony. Accession numbers of the sequences are given in tab. 1.

Distance and parsimony calculations for *Eurychasma* were based on alignments of 45 species with 1746 bases each. Branch support was determined by bootstrap analysis (Felsenstein 1985) using 100 replicates. *Thraustochytrium multirudimentale* was chosen as outgroup.

Fig. 5. Partial alignment of fungal SSU rDNA sequences, including *Chytridium polysiphoniae*, which shows signatures of both the *Rhizophydium* and the *Spizellomyces* clades, distinctly different from those of the two other *Chytridium* species. Details of the presented species can be found in table 2.

Fig. 6. Parsimony tree of *Chytridium polysiphoniae* in the context of the Chytridiomycota. Parsimony calculations for *Chytridium* were done with 38 species and 1034 characters per species, respectively, whilst for Neighbor-Joining (not shown) 35 species and 1027 positions were used. Branch support was determined by bootstrap analysis (Felsenstein 1985) using 100 replicates. *Allomyces macrogynus* was chosen as outgroup for the Chytridiomycetes. Bold figures above the branching are the bootstrap values (100 replicates) for parsimony, figures in italics below are bootstrap values for Neighbor- Joining.

Figure 1

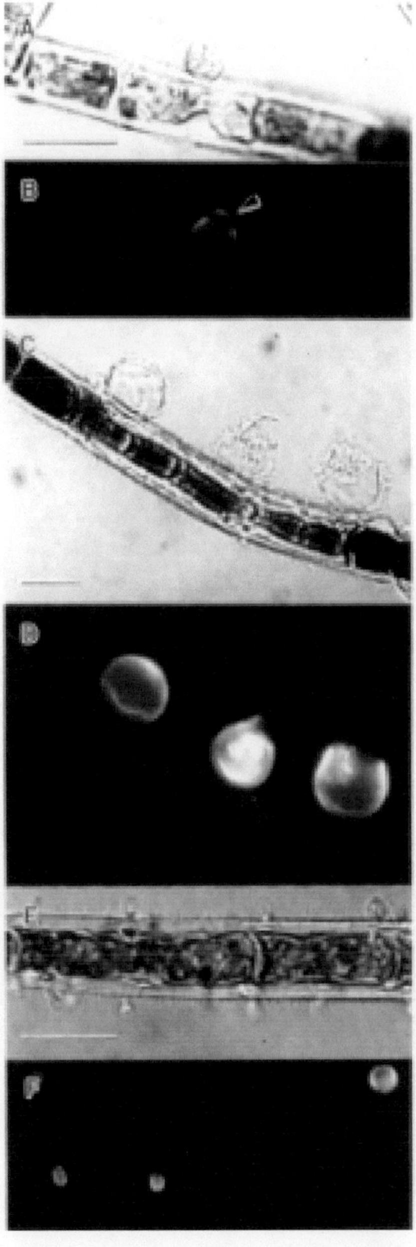

Figure 2

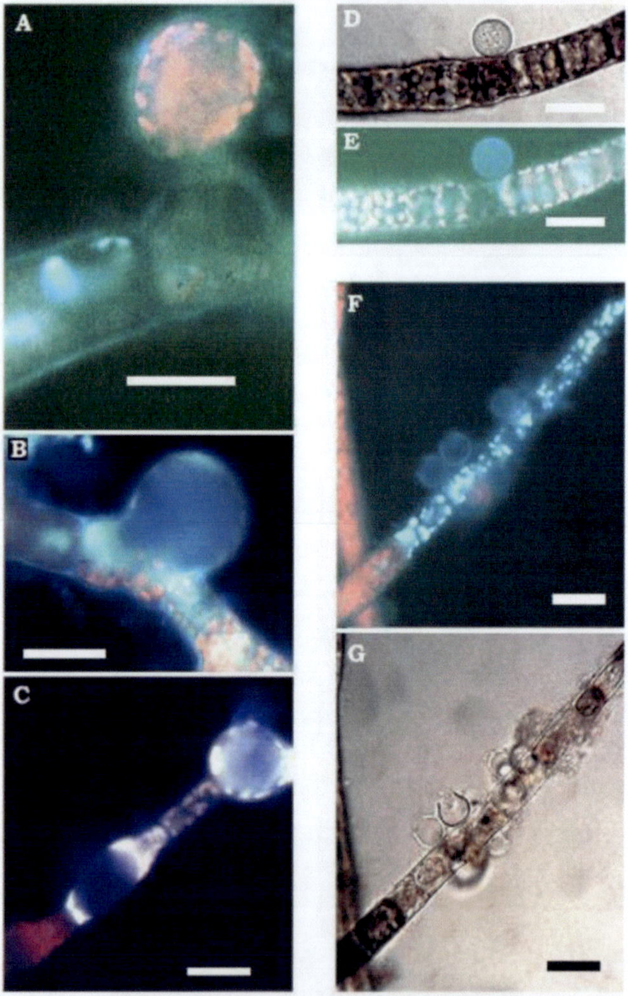

Figure 3

positions 65 - 282 of the SSU rDNA of *Eurychasma dicksonii*
(i.e. from position 69 of the SSU rDNA of *Saccharomyces cerevisiae*, hair pin 6)

```
                  ->       <-
D. eleg.    TAAACTACTTTAAAC 34 AGTTTRTTTG 50 CTTC-AATCCCCGACTGCTCG-CGGAAGGG
H. caten.   TAAACAACTCTATAC 34 AGTTTATTTG 50 CATGAAAT-CCCGACT--TT-T-GGAAGGG
OLI11008    AAAGCAACTTT--AC 34 AGTCTACTTG 50 CAT-AAAAGCCCAACC---TCT-GGACGGG
OLI11026    AAAACAACTTT--AC 34 AGTTTACTTG 50 CATC-AACTCCCAACT---TCT-GGACGGG
E. dicks.   TAAACAATTTTGTAC 34 AGTTTATTTG 50 CAT-AAATACCCGACT--TTTT-GGAAGGG
---------------------------------------------------------------------------
A. brach.   TAAACAATTTTGTAC 34 AGTCTACTTG 50 CGC-AAATACCCAACTGCTTGTCGGACGGG
L. gigan.   TAAACAACTTTGTAC 34 AGTCTACTCG 50 CGC-AAATACCCAACTGCTTGTCGGGCGGG
P. undul.   TAAACACTTTTGTAC 34 AGTCTACTCG 50 CAT-AAATACCCAACTGCTTGTCGGGCGGG
P. monos.   TAAACAATTTTGTAC 34 AGTCTACTCG 50 CAT-AAATACCCAACTGCTTGTCGGGCGGG
P. cymosa   TAAACAATTTTGTAC 34 AGTCTACTTG 50 CGT-AAATACCCAACTGCTTGTCGGACGGG
A. terr.    TAAACAATTTTGTAC 34 AGTCTACTTG 50 CGT-AAATACCCAACTGCTTGTCGGACGGG
S. ferax    TAAACAATTTTGTAC 34 AGTCTACTTG 50 CGT-AAATACCCAACTGCTTGTCGGACGGG
L. chap.    ---------------    ---------- 50 CGT-AAATACCCAACTGCTTGTCGGACGGG
A. bisex.   TAAACAATTTTGTAC 34 AGTCTACTTG 50 CGT-AAATACCCAACTGCTTGTCGGACGGG
A. apic.    TAAACGATTTTGTAC 34 AGTCTACTGG 50 CGT-AAATACCCAACTGCTTGTCGGACGGG
L. chap.1   TAAACAAGTTTGTAC 34 AGTCTACTTG 50 CGT-AAATACCCAACTGCTTGTCGGACGGG
L. cau.     TAAACAAATTTGTAC 34 AGTCTACTTG 50 CGA-AAATACCCAACTGCTTGTCGGACGGG
```

```
D. eleg.    GTGTA 20 GCGGG-TTCGCCCCGTATTGTGGTGAGTC 18 GCATGGCCTTGCGCCGGCGAT
H. caten.   ATGTA 20 GCAGGGGCAACCCTGTATTGTGGTGAGTC 18 GCATGGCCTTGCGCCGGCGAC
OLI11008    CCGCG 20 A---TCTTC-----------TGGTGAATC 18 GCATGGGTTCATCCCGGCGAC
OLI11026    ACGCG 20 GC--TCTTCGGA--GTTTTGTGGTGATTC 18 GCAT-GCGTCA-GC-GGCGAT
E. dicks.   TAGCA 20 GCGGC-TTT--GCCGAATTGAGTTGAATC 18 GCAC----TCT---GTGCGAT
-----------------------------------------------------------------------
A. brach.    AGCA 20 GCAGC-TTCG-GCTGTATTGTGTTGAGTC 18 GCGTT--CTTT--AACGCGAT
L. gigan.   TAGCA 20 GCAGTCTTCGGGCTGTATTGTGTTGAGTC 18 GCAC----TT----GTGCGGT
P. undul.   TAGCA 20 GCAGTCTTCGGGCTGTATTGTGTTGAGTC 18 GCGC---TTTT---GCGCGAT
P. monos.   TAGCA 20 GCAGTCTTCGGGCTGTATTGTGTTGAGTC 18 GCAC----TTT---GTGCGAT
P. cymosa   TAGCA 20 GCGGC-TTCG-GTCGTATTGTGTTGAGTC 18 GC-T----TCAC---AGCGAT
A. terr.    TAGCA 20 GCGGC-CTCG-GTCGTATTGTGTTGAGTC 18 GC-T----TCAC---AGCGAT
S. ferax    TAGCA 20 GCGGC-CTCG-GTCGTATTGTGTTGAATC 18 GC-T----TCAC---AGCGAT
L. chap.    TAGCA 20 GCGTC-TTCG-GGCGTGTTGTGTTGAGTC 18 GC-T----TCAC---AGCGAT
A. bisex.   TAGCA 20 GCGGC-TTCG-GTCGTATTGTGTTGAGTC 18 GC-T---TTT----AGCGAT
A. apic.    TAGCA 20 GCGAC-TTCG-GTGGTGTTGTGTTGAGTC 18 GC-T----TCAC---AGCGAT
L. chap.1   TAGCA 20 GCGCC-TTCG-GGCGTATTGTGTTGAGTC 18 GC-T---TTCG----AGCGAT
L. cau.     TAGCA 20 GCGCT-TTCG-GGCGTATTGTGTTGAGTC 18 GC-T---TTCG----AGCGAT
```

positions 712-825 of the SSU rDNA of *Eurychasma dicksonii*

```
D. eleg.    AGCTGGTCTTAGTTGATCGGTTGTG-GGATGCCCAT 33 CTTATGCC---GC-TG 26 AG
H. caten.   TTCTGGCATTGAGTTGTCGGGGCAG-GGA-ACCCAT 33 CTTACGCC---G-TTG 26 AG
OLI11008    CA-TGGCCTTAATTGGTTGTGGTAT-GGA-ACCCAT 33 CTTAGGCC---G-TTG 26 AG
OLI11026    G-CGGCCGTTAACTCGGTTGTATGTGGGA-ACACAT 33 CTTAGGCC---G-TTG 26 AA
E. dicks.   TACTGCCATTAATTTGGTGGTTTATCAATCTTGCAT 33 CTTTGGCCAAATCTTG 26 AT
-------------------------------------------------------------------------
A. brach.   TTCTGCCATTAAGTTGGTGGTTAGGTGGACTTGCAT 33 CGTTTGCTCA--TTTG 26 AC
L. gigan.   TTCTGCCATTAAGTTGGTGGTTGAATGGACTTGCAT 33 CGTTTGCTCA---TTG 26 AC
P. undul.   TTCTGCCATTAAGTTGGTGGGTTGGTGGGCTTGCAT 33 CGTTTGCTCA--TTTG 26 AC
P. monos.   TTCTGCCATTAAGTTGGTGGTTTGGTGGGCTTGCAT 33 CGTTTGCTCA---TTG 26 AC
P. cymosa   TTCTGCCATTCAGTTGGTGGTTGAGTCGACTTGCAT 33 CGTTTGCTCA--TTTG 26 AC
A. terr.    TTCTGCCATTCAATTGGTGGTTGAGTCGACTTGCAT 33 CGTTTGCTCA--TTTG 26 AC
S. ferax    TTCTGCCATTCAGTTGGTGGTTGTGTCGACTTGCAT 33 CGTTTGCTCA--TTTG 26 AC
L. chap.    TTCTGCCATTGAGTTGGTGGTTGGGTAGACTTGCAT 33 CGTTTGCTCA--TTTG 26 AC
A. bisex.   TTCTGCCATTCAGTTGGTGGTTGAGTAGACTTGCAT 33 CGTTTGCTCA--TTTG 26 AC
A. apic.    TTCTGCCATTGAGTTGGTGGTTGAGTCGACTTGCAT 33 CGTTTGCTCA--TTTG 26 AC
L. chap.1   TTCTGCCATTGAGTTGGTGGTTGGGTAGACTTGCAT 33 CGTTTGCTCA--TTTG 26 AC
L. cau.     TTCTGCCATTCAGTTGGTGGTTGGGTAGACTTGCAT 33 CGTTTGCTCA--TTTG 26 AC
```

Figure 4

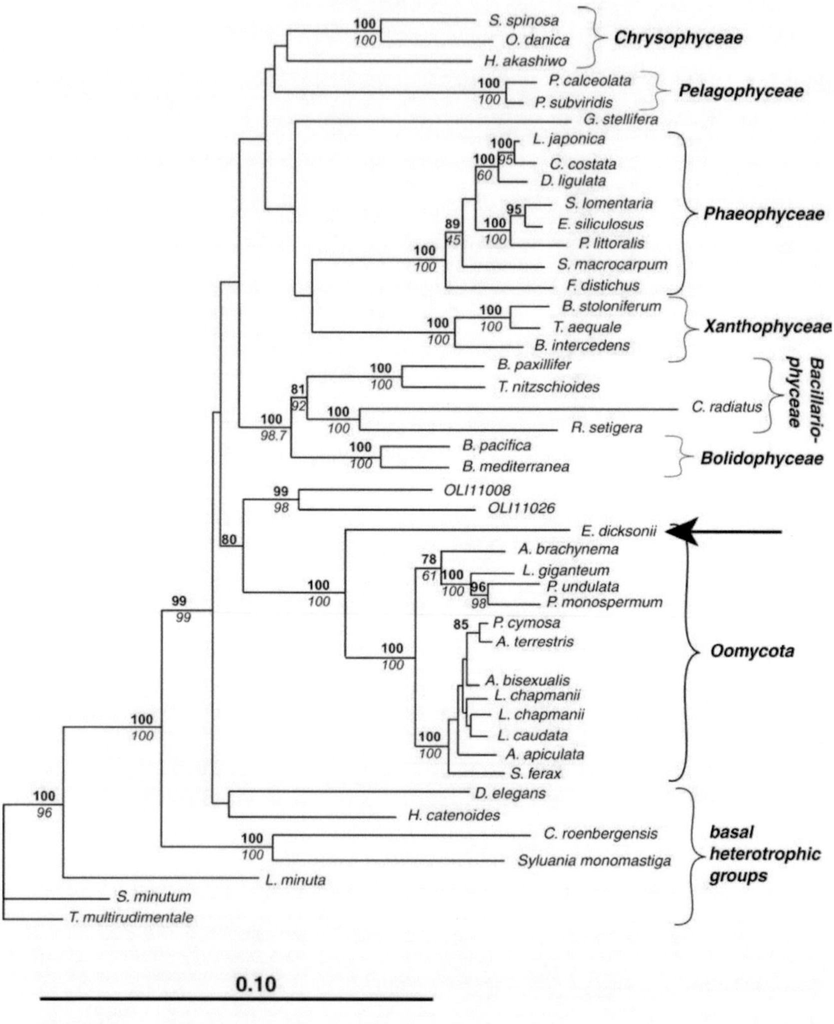
S. spinosa
O. danica
H. akashiwo
Chrysophyceae
P. calceolata
P. subviridis
Pelagophyceae
G. stellifera
L. japonica
C. costata
D. ligulata
S. lomentaria
E. siliculosus
P. littoralis
S. macrocarpum
F. distichus
Phaeophyceae
B. stoloniferum
T. aequale
B. intercedens
Xanthophyceae
B. paxillifer
T. nitzschioides
C. radiatus
R. setigera
Bacillario-phyceae
B. pacifica
B. mediterranea
Bolidophyceae
OLI11008
OLI11026
E. dicksonii
A. brachynema
L. giganteum
P. undulata
P. monospermum
P. cymosa
A. terrestris
A. bisexualis
L. chapmanii
L. chapmanii
L. caudata
A. apiculata
S. ferax
Oomycota
D. elegans
H. catenoides
C. roenbergensis
Syluania monomastiga
L. minuta
basal heterotrophic groups
S. minutum
T. multirudimentale
0.10

Figure 5

positions 66 - 366 of the SSU rDNA of *Chytridium polysiphoniae*
(i.e. from position 69 of the SSU rDNA of *Saccharomyces cerevisiae*, hair pin 6)

```
K. spec.   TAAACAA-TTTT-TACT 45 AGTAAC 40 TGCATTA-AAAACCC 11 GGGTT 22
207 m. rh.TAAACAA--TTG-TACT 45 AGTTCC 40 TGCAA-AAAAATCCC 11 GGGAT 22
L. hiemal.----CAA-TTTT-TACT 45 AGTACC 40 TGCAA-AAAAACCCC 11 GGGGT 22
P. aggreg.---GCAA-TTTT-TACT 45 AGTACC 40 TGCA-CAAAAATCCC 11 GGGAT 22
R. chaet.  TAAGCAC-TT-TATACT 45 AATACC 40 TGCGTCA--AGCCCC 11 GGGGT 22
R. sp.F16  TAAGCAC-TT-TATACT 45 AATACC 40 TGC-TTA-AAATCCC 11 GGGAT 22
R.harder.  TAAGCAA-TT-TATACT 45 AATACC 40 TGC--TAAAAACCTC 11 GAGGT 22
R.sphaer.  TAAGCAC-TT-TATACT 45 AATACC 40 TGC--TAAAAACCCC 11 GGGGT 22
R. sp.138  --AGCAC-TT-TATACT 45 AATACC 40 TGC-TTA-AAATCCC 11 GGGAT 22
C. poly.   TAAACAAATT-TATACT 45 AGTACC 40 TGCAACA--AATCCC 13 GGGAC 22
R. ros.47  TAAACAAATT-TATACT 45 AGTACC 40 TGC--TAAAAATCCC 11 GGGAT 22
R. ros.57  TAAACAATTT-TGTACT 45 AGTACC 40 TGC--TAAAAATCCC 11 GGGAT 22
G. semig.  TAAACAAATT-TGTACT 45 AGTACC 40 TGC--TAAAAATCCC 11 GGGAT 22
S. acum.   TAAACAAATT-TGTACT 45 AGTACC 40 TGC--TAAAAATCCC 11 GGGAT 22
S. kniep.  TAAACAAATT-TGTACT 45 AGTACC 40 TGC--TAAAAATCCC 11 GGGAT 22
P. var.91  TAAACAAATT-TGTACT 45 AATACC 40 TGC--TAAAAATCCC 11 GGGAT 22
P. sp.95   TAAACAAATT-TGTACT 45 AATACC 40 TGC--TAAAAATCCC 11 GGGAT 22
P. hirtus  ----------------- 45 AATACC 40 TGC--TAAAAATCCC 11 GGGAT 22
P. var.85  TAAACAAAATTTTTACT 45 AATACC 40 TGC--TAAAAATCCC 11 GGGAT 22
A. sarc.   TAAACAAATT-TATACA 45 AATACC 40 TGCATTACAAGCCCC 11 GGGGT 22
C. spin.   TAAACAAATTTTGTACT 45 AATACC 40 TGCAAT-CAAACCCC 11 GGGGT 23
O. mucr.   TAAACTCGTT-TGTACT 47 ATAACC 40 TGCAATA-AAATCCT 11 GGGAT 22
P. endog.  TAAACTCGTT-TGTACT 47 ATAACC 40 TGCAATA-AAAGCCT 11 GGGCT 22
R. sp.15   TAAACAAATT-TGTACT 45 AATACC 40 TGCAATA-AAAGGCC 11 GGCCT 22
C. sp.DU   TAAACAAATT-TGTACT 45 AATACC 40 TGCAATA-AAAGGCC 11 GGCCT 22
C. conv.   TAAACAAATT-TGTACT 45 AATACC 40 TGCAATA-AAAGGCC 11 GGCCT 22


K. spec.   ATGGGA-GCAA-TCCTTTTTTTGGTGATTCATAGT 12 CGTATT-GCTTTACGCAGAC 76 GAG
207 m. rh.ATGCC-TTC----GGGTTTACTGGTGATTCATAGT 12 CGTATG-ACTTTATGTCGAC 76 GAG
L. hiemal.ATCCTA-GCAA-TAGGTTTCTTGGTGATTCATAGT 12 CGTATG-ACTTTATGTCGAC 76 GAG
P. aggreg.ATCCCA-GCAA-TGGGTTTCCTGGTGATTCATAGT 12 CGTATG-ACTTTACGTCGAC 76 GAG
R. chaet.  ATCGG--GCAA-CCG-TTTTCTGGTGAATCATGAT 12 CGCATG-ACTCT--GTCGGC 76 GAG
R. sp.F16  ATCGGG-GCAACCCG-TCTACTGGTGAATCATGAT 12 CGCATG-TCATT--GACGGC 76 GAG
R.harder.  ATCGGCTTCG-GCCG-TCTACTGGTGAATCATGAT 12 CGCATG-ACTTT-TGTCGGC 76 GAG
R.sphaer.  ATCG-CTTCG-G-CG-TCTTCTGGTGAATCATGAT 12 CGCATG-GCCTTGCGCCGGC 76 GAG
R. sp.138  ATCGAG-GCAACTCG-TTATTAGGTGAATCATGAT 12 CGTATG-ACTTTACGTCGAC 76 GAG
C. poly.   ACCCGG-GCAA-CCGGATTTTAGGTGATTCATGAT 12 CGTATG-GCTTTGTGCCGAC 76 GAG
R. ros.47  ACCCGG-GCAA-CCGGTTTTTTGGTGATTCATGAT 12 CGTATG-ACCTTGTGTCGAC 76 GAG
R. ros.57  ACCCGG-GCAA-CCGGTTTTTTGGTGATTCATGAT 12 CGTATG-ACCTTGTGTCGAC 76 GAG
G. semig.  ACCCGG-GCAA-CCGGTTTTTTGGTGATTCATGAT 12 CGTATG-GCCTTGTGCCGAC 76 GAG
S. acum.   ACCCGG-GCAA-CCGGTTTTTTGGTGATTCATGAT 12 CGTATG-GCCTTGTGCCGAC 76 GAG
S. kniep.  ACCCGG-GCAA-CCGGTTTTTTGGTGATTCATGAT 12 CGTATG-GCCTTGTGCCGAC 76 GAG
P. var.91  ACCCGG-GCAA-CCGGTTTTTTGGTGATTCATGAT 12 CGTATG-GCCTTGTGCCGAC 76 GAG
P. sp.95   ATCCGG-GCAA-CCGGTTTTTTGGTGATTCATGAT 12 CGTATG-GCCTTGTGCCGAC 76 GAG
P. hirtus  ACCCGG-GCAA-CCGGTTTTTTGGTGATTCATGAT 12 CGTATG-GCCTTGTGCCGAC 76 GAG
P. var.85  ACCCGAGCTTGCTCGGAACTTTGGTGATTCATGAT 12 CGTATGTGCCTTGTGCCGAC 76 GAG
A. sarc.   ACCCG--GCAA--CGGTTCTTTGGTGATTCATGAT 12 CGCATG-GCTTTACGCCGGC 76 GCG
C. spin.   TCCCTGCGCAAGCAGGTTCTCTGGTGATTCATAGT 12 CGCACG-ACCTTGCGTCGGC 76 GCG
O. mucr.   ACCCG--GCAA--CGGTTCTTTGGTGATTCATAGT 12 CGCATG-ACTTTACGTCGGC 76 GCG
P. endog.  ACCCG--GCAA--CGGTTCTTTGGTGATTCATAGT 12 CGCATG-ACTTTACGTCGGC 76 GCG
R. sp.15   ACCCG--GAAA--CGGTTCTTTGGTGATTCATAGT 12 CGCATG-ACTTTACGTCGGC 76 GCG
C. sp.DU   ACCCG--GAAA--CGGTTCTTTGGTGATTCATAGT 12 CGCATG-ACTTTACGTCGGC 76 GCG
C. conv.   ACCCG--GAAA--CGGTTCTTTGGTGATTCATAGT 12 CGCATG-ACTTTACGTCGGC 76 GCG
```

Figure 6

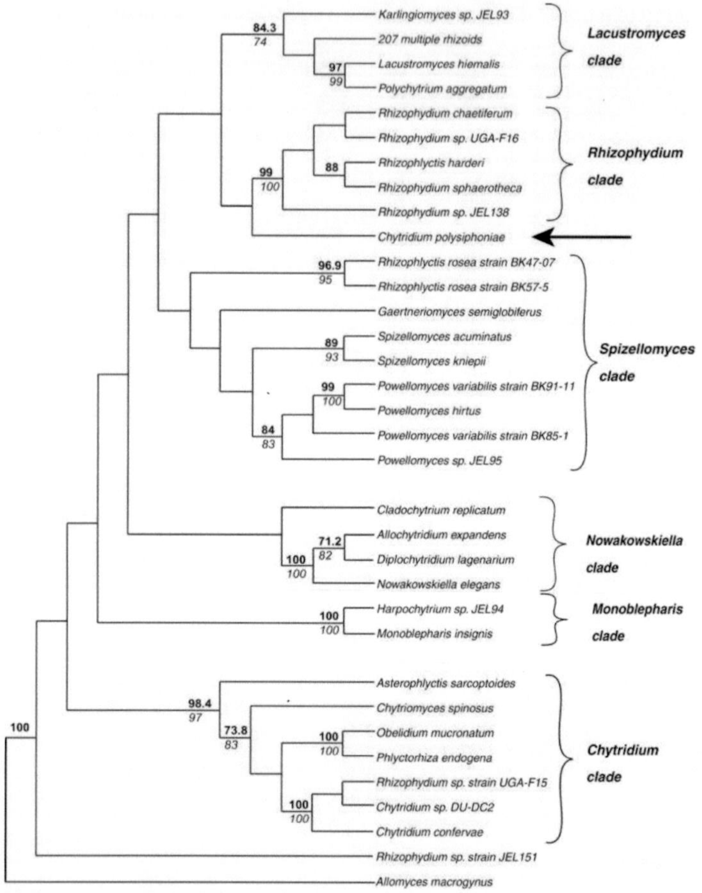
Karlingiomyces sp. JEL93
207 multiple rhizoids
Lacustromyces hiemalis
Polychytrium aggregatum
Lacustromyces clade
Rhizophydium chaetiferum
Rhizophydium sp. UGA-F16
Rhizophlyctis harderi
Rhizophydium sphaerotheca
Rhizophydium sp. JEL138
Rhizophydium clade
Chytridium polysiphoniae
Rhizophlyctis rosea strain BK47-07
Rhizophlyctis rosea strain BK57-5
Gaertneriomyces semiglobiferus
Spizellomyces acuminatus
Spizellomyces kniepii
Powellomyces variabilis strain BK91-11
Powellomyces hirtus
Powellomyces variabilis strain BK85-1
Powellomyces sp. JEL95
Spizellomyces clade
Cladochytrium replicatum
Allochytridium expandens
Diplochytridium lagenarium
Nowakowskiella elegans
Nowakowskiella clade
Harpochytrium sp. JEL94
Monoblepharis insignis
Monoblepharis clade
Asterophlyctis sarcoptoides
Chytriomyces spinosus
Obelidium mucronatum
Phlyctorhiza endogena
Rhizophydium sp. strain UGA-F15
Chytridium sp. DU-DC2
Chytridium confervae
Chytridium clade
Rhizophydium sp. strain JEL151
Allomyces macrogynus
84.3
74
97
99
99
100
88
96.9
95
89
93
99
100
84
83
71.2
82
100
100
100
100
98.4
97
73.8
83
100
100
100
100
100

5

Witnessing effects of pathogens on photosynthesis of their host at the cellular level: two-dimensional chlorophyll fluorescence kinetic measurements of *Pylaiella littoralis* (Phaeophyceae) infected by *Eurychasma* and *Chytridium*

Frithjof C. Küpper[1), 2)], Hendrik Küpper[2), 3)*)], Claire Gachon[4)], Ivan Šetlík[3), 5)]

1) UMR 1931 CNRS-Goëmar, Station Biologique, Place G. Teissier, BP 74, F-29682 Roscoff, Bretagne, France

2) Dept. of Biology, Faculty of Science, D-78457 Konstanz, Germany

3) Microbiological Institute, ASCR, Department of Autotrophic Microorganisms, Opatovický mlýn, CZ-37981 Třeboň

4) Ecole Normale Supérieure, 45, rue d'Ulm, F-75005 Paris, France.
Current address: Institut de Biologie Moléculaire des Plantes, CNRS UPR 406, 12 rue du Général Zimmer, F-67084 Strasbourg cedex, France.

5) Biological Faculty, University of South Bohemia, Branišovská 31, CZ-370 05, České Budějovice, Czech Republic

*) To whom correspondence should be addressed. E-mail: hendrik.kuepper@uni-konstanz.de

Abstract

Spatially resolved measurements of chlorophyll fluorescence induction using fluorescence kinetic microscopy (FKM) were used to study the impact of the parasites *Eurychasma dicksonii* (Oomycota) and *Chytridium polysiphoniae* (Chytridiomycota) on the photosynthesis of their host, *Pylaiella littoralis* (Phaeophyceae). *Chytridium* has a detrimental effect on host photosynthesis, reflected by decreasing F_0 / F_0'. The particularly pronounced decrease of F_0 / F_0' indicates a decreased non-photochemical quenching, while the photochemistry of photosystem II itself (measured as F_V / F_M) remains unaffected until a very late stage of damage to the cell. In contrast, increased values of (F_M – F_s) / F_M and F_0 / F_0' suggest that young infection stages of *Eurychasma* have a stimulatory effect on parts of the photosynthetic apparatus of their algal host, especially on non-photochemical quenching. The latter increases both in amplitude and in speed after the onset of actinic light exposure. In the second half of the infection process, all photosynthetic parameters are on the decline, including the efficiency of photosystem II as measured by F_V/F_M.

Abbreviations

CCD = charge-coupled device, the light detector of digital video cameras; Chl = chlorophyll; FKM = fluorescence kinetic microscope: A microscope allowing two-dimensional (=imaging) records of fluorescence kinetics; F_0 = minimal fluorescence yield of a dark adapted sample, fluorescence in non-actinic measuring light; F_0' = minimal fluorescence yield of a light adapted sample, fluorescence in non-actinic light after actinic irradiance; F_M = maximum fluorescence yield of a dark adapted sample; F_M' = maximum fluorescence yield of a light adapted sample, during the exposure to actinic light, and reduced by non-photochemical quenching; F_p = fluorescence yield at the peak level of the induction curve after the onset of actinic irradiance; F_s = steady state fluorescence under the given actinic irradiance, i.e. after the end of the induction transient; F_v = variable fluorescence; F_v = F_M-F_0 ; LED = light emitting diode; MCP = micro-channel plate, the device in generation 2 and 3 image intensifiers which amplifies the signal; NPOR = net photosynthetic oxygen release; PS II = Photosystem II; q_P = photochemical quenching = (F_M'-F) / (F_M'-F_0') (van Kooten & Snel, 1990); q_{NP} = non-photochemical quenching = 1-q_p (van Kooten and Snel, 1990); rfd = relative fluorescence decrease; Φ_F = fluorescence yield: mol emitted fluorescence photons / mol absorbed excitation photons; $\Phi_{PS\,II}$ = PS II efficiency measured as F_v/F_M

Key words

chlorophyll fluorescence imaging – *Chytridium* - *Eurychasma* – pathogenesis - Phaeophyceae - photosynthesis - *Pylaiella* –

Introduction

All heterotrophic plant pathogens ultimately rely on photosynthesis of their hosts for their own development: whilst biotrophic pathogens and parasites tap their living host's metabolism directly, necrotrophs first kill their host, but still rely indirectly on the matter assimilated by host photosynthesis. The first provide an interesting, but complex system to study the regulation of host metabolism.

Whilst structural and biochemical studies of the localization of nutrient transfer from host plants to their biotrophic parasites have made considerable progress in the last ten years (reviewed by Hall & Williams, 2000), direct measurements of photosynthetic light reactions in the context of pathologies have remained intrinsically difficult. Infections are usually spatially heterogeneous, limiting the applicability of standard methods for photosynthetic measurements such as gas exchange measurements and one-dimensional (single spot) fluorescence kinetic measurements, and producing inconclusive data. So far, the application of spatially resolved methods to host-pathogen interactions has mainly been realized on a macroscopic level, e.g. using chlorophyll fluorescence imaging techniques for rust-infected bean leaves (Peterson & Aylor, 1995) and tobacco leaves infected with tobacco mosaic virus (Balachandran *et al.*, 1994). Scholes and Farrar (1986) used a ^{14}C autoradiographic method to quantify spatial differences of carbon assimilation in rust-infected barley leaves. Following the development of a quantitative imaging technique for chlorophyll fluorescence (Rolfe & Scholes, 1995), Scholes and Rolfe (1996) and Chou *et al.* (2000) carried out a first approach of actually quantifying spatial chlorophyll fluorescence induction on the scale of entire, rust-infected oat and *Arabidopsis* leaves, respectively. However, due to limitations inherent to the technique used (non-modulated excitation), the measurement of intrinsic fluorescence yield F_0 (and thus all parameters related to the photochemical efficiency of PS II) remained impossible. The low sensitivity of the camera used only allowed to take images at saturating light intensity and with a maximum time resolution of 1 s. Furthermore, all macroscopic approaches can be unsatisfactory in certain cases where the first events in infection cycles (taking place on a level of single or few cells) are to be considered, and unsuitable in the case of holocarpic pathogens, i.e. in cases where a single pathogen spore develops into a unicellular infection structure affecting only a single or very few host cells, later converting itself into a new sporangium. Because of these technical limitations, plant-pathogen interactions have been much better studied in aspects of secondary metabolism than in their bioenergetical aspects.

Regarding pathologies of marine plants, still fewer studies have considered photosynthesis. Durako and Kuss (1994) reported impaired photosynthesis (measured as net photosynthetic oxygen release, NPOR) in the seagrass *Thalassia testudinum* upon infection by the slime mould *Labyrinthula schizochytrops*, contributing to the die-off of seagrass beds off Florida. Goff (1979, 1982) observed translocation of ^{14}C-labelled photoassimilates from red algal hosts to parasitic red algae with little or no photosynthetic pigments. In the case of virus infections of filamentous Ectocarpalean brown algae, results obtained with macroscopic measurements of NPOR appear contradictory: Whilst Robledo *et al.* (1994) had observed a reduction of photosynthetic capacity in virus-infected cultures of *Feldmannia*, del Campo *et al.* (1997) observed no significant difference in healthy and virus-infected *Ectocarpus siliculosus*.

However, Ectocarpales are plagued by a variety of pathogens: Müller *et al.* (1998) mention a number of virus infections, Maier *et al.* (2000) described *Maullinia ectocarpi* as a novel plasmodiophoralean pathogen of *Ectocarpus siliculosus*, and Küpper and Müller (1999 / Chapter 2) reported frequent epidemics of the heterokont and fungal pathogens *Eurychasma dicksonii*, *Chytridium polysiphoniae* and *Anisolpidium rosenvingei* in European populations of *Pylaiella littoralis*. Culture studies revealed that *Eurychasma dicksonii* and *Chytridium polysiphoniae* have large host ranges comprising members of most orders of brown algae. A complete infection cycle of *Eurychasma* in *Pylaiella* has a duration of 16 days minimum, whilst *Chytridium* takes only 6 days for a complete infection cycle in this host (Müller *et al.*, 1999 / Chapter 3). Recent molecular studies clarified the phylogenetic position of these two pathogens (Küpper *et al.*, 2001a / Chapter 4). Both pathogens are holocarpic, i.e. a spore infects a single host cell and subsequently develops into a new pathogen sporangium, without formation of hyphae and infection of neighbouring cells. In the case of *Eurychasma*, the pathogen cytoplasm enters the host cell, which subsequently shows hypertrophic growth, strongly increasing its volume before being filled entirely by the developing pathogen sporangium (Küpper & Müller, 1999 / Chapter 2). In contrast, the developing *Chytridium* sporangium remains on the surface of the host cell, only inserting a haustorium-like structure into it, in order to recover nutrients from the host.

Until close to the end of the infection cycle, *Eurychasma*-infected *Pylaiella* cells still exhibit chlorophyll autofluorescence (Küpper *et al.*, 2001a / Chapter 4). Together with the hypertrophic growth of infected cells, this indicates that host photosynthesis continues. In contrast, *Chytridium*-infected cells lose chlorophyll autofluorescence within a few days after infection and show accumulation of blue-autofluorescent compounds, possibly polyphenolics. These two parasites could therefore be interesting models for studying the impact of infection on host photosynthesis in two very different types of interaction. It is obvious that in the case of such infections, macroscopic approaches are completely unsuitable to provide any insight, since infected cells are interspersed between uninfected ones and because they are likely to react differently depending on the developmental stage of the infection.

Chlorophyll fluorescence measurements are a reliable, non-invasive, quantitative tool to assay the function of the photosynthetic apparatus. It enables the study of parameters like the photochemical and non-photochemical energy dissipation, organization of light harvesting systems, etc. (reviewed by Krause & Weis 1991; Dau 1994; Govindjee 1995; Lazár, 1999). The potential of the method has been further stimulated in the last decade by the introduction of the imaging detectors (CCD cameras and MCP image intensifiers) which provide a time sequence of two-dimensional maps of fluorescence emission. Because of limited sensitivity and speed, early imaging systems were only able to record images of the maximum and steady state levels of the fluorescence induction (e.g. Omasa *et al.* 1987, Genty & Meyer, 1994; Rolfe & Scholes, 1995; Siebke & Weis, 1995). Oxborough & Baker (1997) could record F_0 images, but only using extremely long (15 min) integration times and an external irradiance measurement to calculate the quantum yield. The assessment of fluorescence quantum yield under widely different irradiances is best performed by measuring the fluorescence response to short (tens of μs) measuring light pulses of standard strength applied at an appropriate frequency (Bilger & Schreiber 1990; Schreiber *et al.* 1993, 1995; Govindjee 1995). Only very recently this principle was successfully introduced to imaging measurements of chlorophyll fluorescence induction (Nedbal *et al.* (2000) on the macroscopic scale; Küpper *et al.* (2001b) on the microscopic scale). This measuring principle allows the recording of complete two-dimensional kinetics of fluorescence quantum yield with high time resolution (40 ms).

In phytopathology in general and in the case of *Eurychasma*- and *Chytridium*-infected *Pylaiella* in particular, this novel technique offers an interesting perspective to unveil the effects of infections on the primary events of host photosynthesis. The fluorescence kinetic microscope developed by Küpper *et al.* (2001b) allows measurements of photosynthetic performance at the cellular and sub-cellular level, for the first time with all the advantages inherent to the technique with pulsed measuring light. Designed to detect spatial differences in chlorophyll fluorescence induction kinetics down to the size of individual chloroplasts, it is generally suitable to study cases of stress in which single photosynthetically active cells (or individual cells in a tissue) are likely to react differently.

Materials and methods

Biological material

Infected *P. littoralis* was collected at Aith Voe (Bressay, Shetland) in April 1996 (Küpper & Müller, 1999). Unialgal host cultures were established as described previously (Müller *et al.*, 1999) using a unialgal clonal culture of *P. littoralis* (Pyl IR) from Isla Diego Ramirez (Chile, Drake Passage; Müller & Stache, 1989), yielding the unialgal host cultures Eu Pil IR 1 starting from spores of one *E. dicksonii* sporangium and Chyt Pyl IR 14 starting from spores of 14 sporangia of *C. polysiphoniae*. These host-parasite associations were used for all our experiments.

Cultures were maintained in plastic Petri dishes in Provasoli ES medium (Starr & Zeikus, 1987) prepared from natural seawater from the Atlantic off Roscoff, Brittany. They were illuminated with daylight-type fluorescent lamps at an irradiance of 9 $\mu E\ m^{-2}\ s^{-1}$ for 10h day^{-1} and kept at 10 ±1°C. Cultures were transferred to fresh medium in one or two week intervals.

Fluorescence kinetic microscopy

Single-cell measurements of chlorophyll fluorescence induction were carried out using the system recently described by Küpper *et al.* (2001b). In this system, four light sources are used during the measurement: pulsed measuring light, actinic light, saturating flashes, and transmittant observation light.

In all measurements of *Pylaiella*, the flashes of non-actinic measuring light were produced by a blue LED (Nichia, Tokushima, Japan: NSPB500S, λ_{max}=470 nm, optical output power 6 mW). The duration of these measuring flashes, the gating of the MCP image intensifier and the shutter of the camera (both synchronized with the flashes) were either set to 30 µs or 100 µs integration time.

Both actinic and saturating light were passed through a set of filters providing a blue excitation light reaching from 380 nm to 450 nm. When performing light saturation measurements (PI curves) at variable light intensities, actinic irradiance was attenuated by neutral density filters to 0-5000 $\mu mol.m^{-2}.s^{-1}$, saturating flashes were attenuated to 1000-2000 $\mu mol.m^{-2}.s^{-1}$. Transmittant observation light was attenuated to about 1 $\mu mol.m^{-2}.s^{-1}$ so that it was not or only weakly actinic, leaving cells in their dark-adapted state.

Gas and temperature conditions during the measurements were controlled using the small-window version of the microscope measuring chamber of Küpper *et al.* (2001b), perfused with air-saturated seawater at 12 °C. Measurements always lasted 50 seconds: during the first second a short (600 ms) flash of saturating light was given for measurement of F_M, followed by 5 seconds with neither actinic nor saturating light for measurement of F_0.

Then, 40 s of actinic light were applied to analyze the Kautsky induction, and finally again 5 seconds with neither actinic nor saturating light for measurement of F_0'. During the actinic light exposure, further 600 ms flashes of saturating light were given in order to measure the photochemical *versus* non-photochemical quenching (q_p, q_{np}).

Data analysis

Data from fluorescence kinetic microscopy were analysed by FluorCam™ software (Photon Systems Instruments / PSI, Brno, Czech Republic) in two ways. On the one hand, images of photosynthetic parameters were obtained by averaging a few frames of the fluorescence kinetic film over the whole recorded field; e.g. for an F_0 image all frames of the fluorescence kinetic film which were within the recording time of F_0 were averaged into one image. To calculate an image of a derived parameter like F_V/F_M, mathematical operations (e.g. subtraction, division) were performed on a pair of images of primary parameters (i.e. F_M, F_0, ...). On the other hand, objects could be designated manually on a frame (or an average of frames) to calculate their fluorescence kinetics by averaging the kinetics of their individual pixels. These curves were exported from the FluorCam™ software in ASCII format and further processed with a self-written add-in program for Microsoft Excel™, which automatically calculated the numerical values and standard errors of all customary photosynthetic parameters. Besides that, it performed an optimized background correction on the measured curves. Relative slopes of the Kautsky kinetics were obtained by smoothing and differentiation using Microcal Origin™.

Size / age classification of infections

Due to the marked morphological changes occurring in infected *Pylaiella* cells in the course of the infection cycles of both *Eurychasma* and *Chytridium*, it appeared essential to introduce a classification for the age of a given infection. In the case of *Eurychasma*, three slightly overlapping categories were distinguished: *Young* stages, corresponding to a ratio of the maximum diameter of an infected *Pylaiella* cell to the diameter of a neighbouring uninfected cell of 1.44 ± 0.25 (Fig. 1 A), corresponding to an age of up to around 5 days; *medium* stages, corresponding to a ratio of the maximum diameter of an infected *Pylaiella* cell to the diameter of a neighbouring uninfected cell of 2.07 ± 0.44 and an age of up to 10-12 days (Fig. 1 B); *late* stages, corresponding to a diameter ratio of 2.80 ± 0.45 and an age of over 10 days up to final spore release (Fig. 1 C).

For *Chytridium*, only two categories were distinguished, subsequently termed *young* and *old* infections: 1. *Young* infections had a ratio of diameters of the *Chytridium* sporangium to that of the corresponding, infected *Pylaiella* cell of up to 0.62 ± 0.26, corresponding to an infection age of up to around 5 days), with the infected *Pylaiella* cells usually still containing visible photosynthetic pigments (Fig. 1 D). 2. O*ld* infections had a ratio of diameters of the *Chytridium* sporangium to that of the corresponding, infected *Pylaiella* cell of at least around 0.8, corresponding to an infection age of at least 5 days, up to spore release), with the infected *Pylaiella* cells often having lost all visible photosynthetic pigments (not shown, since no fluorescence measurements were possible on these cells).

Since this classification is still partly based on arbitrary criteria, it is intrinsically impossible to apply statistical tests on the related data. Error bars in the corresponding figures (Figs 2, 3) hence express variability, expressed as standard deviation.

Results and discussion

Fluorescence parameters of *Pylaiella* infected by *Eurychasma*

The relative variable fluorescence $\mathbf{F_V / F_M}$ is a well-established parameter of the maximum photochemical efficiency of PS II in a dark-adapted sample, the value of which is around 0.83 in unstressed tissues of terrestrial plants (Björkman & Demmig, 1987). F_V / F_M in *Pylaiella littoralis* uninfected cells (Fig. 2) is considerably lower (0.60), but comparable to values measured in other brown algae (Dring *et al.*, 1996; Pearson *et al.*, 2000). It remains stable in young infection stages, suggesting that activity of PS II continues almost normally. However, it is decreasing further with continuing infection in medium stages, indicating a progressive damage to the primary photochemistry of PS II the longer the infection is present.

In contrast, the decrease of fluorescence quantum yield under actinic irradiance $\mathbf{(F_M - F_S)}$ has been established as a more general but sensitive vitality index, introduced by Lichtenthaler and Rinderle (1988) as "relative fluorescence decrease" (rfd). Whilst these authors normalized $F_M - F_S$ to F_S, it was normalized to F_M in the present study because measurements of the latter were more accurate. Compared to uninfected *Pylaiella* cells, it increased in *young* stages of *Eurychasma* infections (Fig. 2). This can be interpreted as an enhanced metabolic activity in the initial stage of *Eurychasma* infection. This might provide additional resources required for the hypertrophic growth of infected cells, which will multiply their volume during the infection cycle. In medium stages, the parameter $(F_M - F_S) / F_M$ is about equal to uninfected cells, and is considerably reduced in late stages. Medium stages may be considered as a break-even point where photosynthetic and metabolic activity have already been reduced from the maximum levels earlier, back to a level corresponding to uninfected cells, on a downward slope of decreasing vitality. However, since all measurements used for this parameter in the present study were done under saturating irradiance, it also reflects increased non-photochemical quenching, which becomes more obvious when considering the change of the intrinsic fluorescence yield due to actinic light exposure, i.e. $\mathbf{F_0 / F_0'}$. Increased values of this parameter reflect non-photochemical quenching. This can have many reasons, most of which are related either to photoinhibition or to adaptation to high irradiances, such as changes in pigment organisation like the xanthophyll cycle (e.g. Baker 1991, Muller *et al.*, 2001, Young *et al.*, 1996). Still, the factors contributing to q_{NP} in general are not fully understood so far, especially in brown algae which are much less studied than green plants or diatoms. Nevertheless, its variations are likely related to modifications of the redox state of the thylakoid membrane: e.g., ascorbate is both a well-known quencher of chlorophyll fluorescence (Ivanov & Edwards, 2000), and is also involved in redox homeostasis during pathogenesis-related oxidative stress in higher plants (Noctor & Foyer, 1998). Several studies concerning the effects of abiotic stresses on photosynthesis of brown algae and diatoms showed the involvement of chlororespiration and the xanthophyll cycle during photoinhibition (Hanelt *et al.*, 1997; Arsalane *et al.*, 1994), and pointed out the xanthophyll cycle as a possible mechanism explaining the "energy-dependant" and "photoinhibitory" components of q_{NP} (reviewed by Krause & Weis, 1991). This parameter increases from uninfected over young to late stages of *Eurychasma* infection, which could suggest an increasing susceptibility to photoinhibition. The latter would indicate increasing low-light adaptation, which should also be obvious from the light saturation (PI) curves. Since this is not the case, it can be concluded that it is caused by different reasons, the nature of which has to be the subject of further studies. Because F_0 / F_0' increases already from non-infected cells to early stages of sporangium development, it may also be the reason for the increased (F_M - F_S) / F_M in this stage of parasite development: a constant photochemical activity (F_V / F_M) together with an increase in non-photochemical quenching (F_0 / F_0') would exactly yield the effect on (F_M - F_S) / F_M that was observed. In further studies, more measurements at non-saturating irradiances will be carried out, which will allow a more detailed analysis of the proportions of photochemical vs. non-photochemical quenching and thus reveal the true reason of the changes of (F_M - F_S) / F_M.

Contrary to the above parameters, $\mathbf{F_p / F_0}$ as a function of irradiance reflects the light-saturation curve of the PS II photochemistry (Fig. 4). The values obtained with *Eurychasma*-infected cells are consistently lower than for uninfected cells, pointing at an increased adaptation to weak light intensities. The very fast decrease of F_p / F_0 at over-saturating irradiances is something very unusual. It suggests that in *Pylaiella* (and possibly also other brown algae) the concept of saturating light pulses for F_M and quenching analysis may not be applicable in the usual sense, i.e. the over-saturating pulses used for higher plants (several thousand $\mu mol.s^{-1}.m^{-2}$) would lead to an underestimation of F_M. A correct F_M of *Pylaiella* would thus only be measured with just saturating irradiance. An alternative explanation could be that the dark adaptation between the high irradiance measurements was not long enough, which would lead to the same effect, without far-reaching conclusions. Nevertheless, this is unlikely to be the main or only reason, since the short over-saturating pulses of light during individual measurements at over-saturating irradiance often led to an immediate further decrease of fluorescence quantum yield and thus to negative values for photochemical quenching, calculated according to van Kooten and Snel (1990). Indeed, the most probable explanation for this phenomenon is a very fast (i.e. in the ms range) increase of non-photochemical quenching upon over-saturating irradiance.

Using a temperature-controlled measuring chamber constantly perfused with PES medium to maintain a single infected *Pylaiella* filament alive in an unchanged, repeatedly observable position, F_p / F_0 was measured in an individual *Eurychasma*-infected cell (belonging to the *late* age group) on two consecutive days. F_p / F_0 was already lower in the infected than in a neighbouring uninfected cell at the beginning, and decreased even further within one day (data not shown). At last, the infected cell died earlier than the surrounding uninfected ones.

Fluorescence parameters of *Pylaiella* infected by *Chytridium*

Both variable fluorescence, F_V / F_M, and (F_M – F_S) / F_M are almost unchanged in infected cells (0.43 compared to 0.44, and 0.60 to 0.56, in uninfected and infected cells of the same filaments, respectively, Fig. 3), indicating that the efficiency of PS II itself is not affected until a late stage of parasite sporangium development. F_0 / F_0' shows a strong decrease in infected cells (Fig. 3). Absolute values for F_M in infected cells were always lower than for uninfected cells (not shown). These parameters consistently show an overall decline in photosynthetic performance due to *Chytridium* infection.

Slopes of Kautsky curves of *Pylaiella* infected by *Eurychasma* and *Chytridium*

Under the present conditions, i.e. saturating light, this parameter reflects the non-photochemical quenching (q_{NP}, because by definition, photochemistry is saturated at an irradiance which saturates F_P / F_0). In contrast to the F_0 / F_0' and (F_M - F_S) / F_M, it reveals the time course of q_{NP}. Both in *Eurychasma*-infected and uninfected cells, the point of maximum slope shifts to shorter times after the onset of actinic light with increasing irradiance. This is accompanied by a change of shape of the Kautsky curve from a sigmoidal to an exponential decrease.

In *Eurychasma*-infected cells (Figs 5 A, B; Fig. 6), Kautsky curves of the early infection stages usually showed steeper maximum slope compared to uninfected cells. In the few cases where this did not occur, a very steep slope at the beginning of the curve was observed instead. These observations show that not only the total extent, but also the speed of non-photochemical quenching is increased in early stages of *Eurychasma* infections. To our knowledge, such differences in curvature of Kautsky curves have not been published before.

In contrast, the slope of Kautsky curves obtained with *Chytridium*-infected was not significantly altered (Fig. 5 C; the oscillations of the slope in the later infection stages visible in are most likely caused by the noise due to the weak measuring signal).

Conclusions

Overall, this study points out a promising, new field of use of the fluorescence kinetic microscopy technique (Küpper *et al.*, 2001b) as a tool to explore the considerable spatial heterogeneities, caused by a complete reprogramming of host metabolism in infected tissues (Chou *et al.*, 2000). One considerable advantage of the technique applied here over current, parallel developments (Scholes, Farrar & Rolfe, unpublished) is the use of fast-pulsed measuring light. This allows direct measurements of fluorescence quantum yield and high (presently millisecond-range) time resolution of imaging chlorophyll fluorescence kinetic recordings (Fig. 6).

The present study suggests an entirely different impact of two pathogens, *Eurychasma dicksonii* and *Chytridium polysiphoniae*, on photosynthesis of their brown algal host *Pylaiella littoralis*. Chlorophyll fluorescence parameters indicate that early infection stages of *Eurychasma* initially enhance certain parameters of the photosynthetic performance of the infected host cell, before turning to a decline in the second half of the infection cycle. These observations imply a subtle tuning of host metabolism towards enhanced photosynthetic performance at the onset of hypertrophic growth, most likely to cover the anabolic needs of the latter. Once these are fulfilled and the developing *Eurychasma* cell is approaching its maximum size, completely filling the host cell, cell organelles of the host are being pressed to the cell periphery and destroyed so that photosynthesis ceases. These observations of early *Eurychasma* infection stages in *Pylaiella* are reminiscent of stimulated photosynthesis in a terrestrial plant-pathosystem, the "green islands" of rust (*Puccinia* sp.)-infected barley and oat leaves, respectively (Scholes & Farrar, 1986; Scholes & Rolfe, 1996). In contrast to *Eurychasma*, *Chytridium* has an immediate, negative impact on host photosynthesis which rapidly ceases within a week after infection. This confirms earlier, observations by bright-field (Chapter 2 / Müller *et al.*, 1999) and UV/fluorescence microscopy (Chapter 3) showing a rapid destruction of cellular structures and accumulation of apparently senescence-related, UV-autofluorescent compounds, respectively. Among higher plants, reports of impaired photosynthesis due to pathogen infections (e.g. Roberts & Walters, 1988; Raggi, 1995; Scholes & Rolfe, 1996; Abo-Foul et al., 1996; Chou *et al.*, 2000) seem to be more numerous than cases of stimulated photosynthesis (e.g. Scholes & Farrar, 1986; Yurina *et al.*, 1996). This reflects the different ecological strategies employed by pathogens, either making use of energy resources of the host over a longer time span, or killing the host very rapidly and showing a drastic pathogenicity.

Still, the study as it is presented here has to be considered preliminary. In a follow-up FKM project on *Eurychasma*- and *Chytridium*-infected *Pylaiella*, it would be necessary to follow the changes of all fluorescence parameters in selected *Eurychasma* and *Chytridium* sporangia over longer time scales than only 2 days. This would result in a true timescale of the development of the various photosynthetic parameters, which would also allow more advanced statistical treatment of the data. At the same time, the effect of host cell age on the development of the parasite (and corresponding changes in photosynthesis) could be studied with this approach. Even though the currently used measuring chamber was permanently perfused with culture medium, the dying of filaments after several days might be due to release of toxic metals from the stainless steel walls because of the highly corrosive properties of sea water. For an optimisation of the measuring chamber for studies of marine organisms in general, a construction consisting entirely of synthetic polymers should be used. Besides that, it would be essential to record light saturation curves extending to much lower actinic light intensities than those applied here, in particular to explore the increased non-photochemical quenching of *Eurychasma*-infected cells.

Furthermore, also studies of changes in PS II antenna size and further important photosynthetic parameters (such as Q_A re-oxidation kinetics) should be carried out to for a more accurate determination of the changes in PS II function resulting from *Eurychasma* and *Chytridium* infections. Inevitably, the correlation between q_{NP} and the slope and shape of Kautsky curves, one of the most intriguing effects of *Eurychasma* infection, will have to be addressed.

In the case of *Eurychasma*-infected *Pylaiella*, FKM measurements combined with a pharmacological approach using specific inhibitors could give hints about the underlying mechanisms causing the described variations of q_{NP}: For example, the use of diphenylene iodonium (DPI, a flavoprotein inhibitor inactivating the production of reactive oxygen species during the oxidative burst in animals, green plants, and brown algae; Küpper *et al.*, 2001c / Chapter 6) or reducing agents such as ascorbate (Noctor & Foyer, 1998) could allow to assess the impact of pathogenesis-related oxidative stress on photosynthetic parameters. Furthermore, the role of chlororespiration should be addressed using myxothiazol or antimycin A, that have been shown to inhibit this process in *Chlamydomonas reinhardtii* and *Phaeodactylum tricornutum*, probably by acting on a cytochrome-containing component of the plastidial electron transport chain (Ravenel & Peltier, 1991 and Caron *et al.*, 1987, respectively).

Still, a future study on this pathosystem should not remain entirely descriptive, regarding fluorescence parameters. It should ultimately address the underlying mechanisms for increased photosynthetic and anabolic activities in *Eurychasma*-infected cells. Concerning these aspects, little but speculations exist. In analogous cases in terrestrial plant-pathosystems, cytokinines were proposed to be the active principles released by plant pathogens, stimulating photosynthesis (Meidner, 1967) by increased protein synthesis (Kende, 1971). In one of the first studies regarding molecular aspects of pathogen impact on photosynthetic light reactions, white blister rust (*Albugo candidans*) infection in *Arabidopsis thaliana* was found to result in inhibited photosynthesis in infected leaf areas, with photosynthesis-related genes being repressed in the concerned tissues.

Acknowledgements

We are grateful to Dieter G. Müller (University of Konstanz) for providing cultures of *Eurychasma dicksonii* and *Chytridium polysiphoniae*. We would like to thank Martin Trtilek (Photon Systems Instruments, Brno, Czech Republic) for providing tailored electronic equipment, adapted for the particular needs of this study. We also wish to acknowledge Todor Macho for carrying out all precision mechanical works during the construction of the FKM including the microscope measuring chamber.. F. C. K. was supported with fellowships from the European Commission (MAST-III Program) and Studienstiftung des Deutschen Volkes (Bonn). H. K. received a scholarship from Studienstiftung des Deutschen Volkes.

References

Abo-Foul S, Raskin VI, Sztejnberg, Marder JB, 1996: Disruption of chlorophyll organization and function in powdery mildew-diseased cucumber leaves and its control by the hyperparasite *Ampelomyces quisqualis*.- Phytopathology 86 (2), 195-9

Arsalane W, Rousseau B, Duval J-C, 1994: Influence of the pool size of the xanthophyll cycle on the effects of light stress in a diatom: competition between photoprotection and photoinhibition.- Photochemistry and Photobiology 60(3), 237-43

Baker NR, 1991: A possible role for photosystem-II in environmental perturbations of photosynthesis.- Physiologia Plantarum 81, (4) 563-70

Balachandran S, Osmond CB, Daley PF, 1994: Diagnosis of the earliest strain-specific interactions between tobacco mosaic virus and chloroplasts of tobacco leaves in vivo by means of chlorophyll fluorescence imaging.- Plant Physiology 104, 1059-65

Bilger W, Schreiber U, 1990: Chlorophyll luminescence as an indicator of stress-induced damage to the photosynthetic apparatus. Effects of heat stress in isolated chloroplasts.- Photosynthesis Research 25, 161-70

Björkman O, Demmig B, 1987: Photon yield of O_2 evolution and chlorophyll fluorescence characteristics at 77 K among vascular plants of diverse origins.- Planta 170, 489-504

Caron L, Berkaloff C, Duval J-C, Jupin H, 1987: Chlorophyll fluorescence transients from the diatom *Phaeodactylum tricornutum*: relative rates of cyclic phosphorylation and chlororespiration.- Photosynthesis Research 11, 131-9

Chou H-M, Bundock N, Rolfe SA, Scholes JD, 2000: Infection of *Arabidopsis thaliana* leaves with *Albugo candida* (white blister rust) causes a reprogramming of host metabolism.- Molecular Plant Pathology 1(2), 99-113

Dau H, 1994: Molecular mechanisms and quantitative models of variable photosystem II fluorescence.- Photochemistry & Photobiology 60, 1-23

del Campo E, Ramazanov Z, García-Reina G, Müller DG, 1997: Photosynthetic responses and growth performance of virus-infected and noninfected *Ectocarpus siliculosus* (Phaeophyceae).- Phycologia 36 (3), 186-9

Dring MJ, Makarov V, Schoschina E, Lorenz M, Lüning K, 1996: Influence of ultraviolet-radiation on chlorophyll fluorescence and growth in different life-history stages of three species of *Laminaria* (Phaeophyta).- Marine Biology 126, 183-91

Durako MJ, Kuss KM, 1994: Effects of *Labyrinthula* infection on the photosynthetic capacity of *Thalassia testudinum*.- Bulletin of Marine Science 54(3), 727-32

Genty B, Meyer S, 1994: Quantitative mapping of leaf photosynthesis using chlorophyll fluorescence imaging.- Australian Journal of Plant Physiology 22, 277-84

Goff LJ, 1979: The biology of *Harveyella mirabilis* (Cryptonemiales, Rhodophyceae). VI. Translocation of photoassimilated ^{14}C.- Journal of Phycology 15, 82-7

Goff LJ, 1982: The biology of parasitic red algae.- Chapter 8 in: Progress in Phycological Research 1 (Round FE and Chapman DJ, eds), 289-369. Elsevier Biomedical Press B.V.

Govindjee, 1995: Sixty-three years since Kautsky: chlorophyll a fluorescence.- Aust. J. Plant Physiol. 22, 131-60

Hall JL, Williams LE, 2000: Assimilate transport and partitioning in fungal biotrophic interactions.- Aust. J. Plant Physiol. 27, 549-60

Hanelt D, Wiencke C, Karsten U, Nultsch W, 1997: Photoinhibition and recovery after high light stress in different developmental and life-history stages of *Laminaria saccharina* (Phaeophyta).- Journal of Phycology 33, 387-95

Ivanov B, Edwards G, 2000: Influence of ascorbate and the Mehler peroxidase reaction on non-photochemical quenching of chlorophyll fluorescence in maize mesophyll chloroplasts.- Planta 210, 765-74

Kende H, 1971: The cytokinins.- International Review of Cytology 31, 301-38

Krause G, Weis E, 1991: Chlorophyll fluorescence and photosynthesis: the basics.- Annual Review of Plant Physiology & Plant Molecular Biology 42, 313-49

Küpper FC, Müller DG, 1999: Massive occurrence of the heterokont and fungal parasites *Anisolpidium*, *Eurychasma* and *Chytridium* in *Pylaiella littoralis* (Ectocarpales, Phaeophyceae).- Nova Hedwigia 69, 381-9 (Chapter 1 of this thesis)

Küpper FC, Maier I, Müller DG, Guillou L, Loiseaux-de Goer S, 2001a: Phylogenetic affinities of two eukaryotic pathogens of marine macroalgae, *Eurychasma dicksonii* and *Chytridium polysiphoniae*.- (*submitted* / Chapter 3 of this thesis)

Küpper H, Šetlík I, Trtilek M, Nedbal L, 2001b: A microscope for two-dimensional measurements of *in vivo* chlorophyll fluorescence kinetics using pulsed measuring light, continuous actinic light and saturating flashes.- Photosynthetica 38(4), *in press*

Küpper FC, Kloareg B, Guern J, Potin, P, 2001c: Oligoguluronates elicit an oxidative burst in the brown algal kelp *Laminaria digitata*.- Plant Physiology 125, 278-291 (Chapter 6 of this thesis)

Lazár D, 1999: Chlorophyll a fluorescence induction.- Biochimica et Biophysica Acta 1412, 1-28

Lichtenthaler HK, Rinderle U. 1988. The role of chlorophyll fluorescence in the detection of stress conditions in plants.- CRC Crit. Rev. Analyt. Chem. 19, Suppl. I, 29-85

Meidner H (1967) The effect of kinetin on stomatal opening and the rate of uptake of carbon dioxide in mature primary leaves of barley.- Journal of Experimental Botany 18, 556-61

Müller DG, Kapp M, Knippers R, 1998: Viruses in marine brown algae.- Advances in Virus Research 50, 49-67

Müller DG, Küpper FC, Küpper H, 1999: Infection experiments reveal broad host ranges of *Eurychasma dicksonii* (Oomycota) and *Chytridium polysiphoniae* (Chytridiomycota), two eukaryotic parasites in marine brown algae (Phaeophyceae).- Phycological Research 47, 217-23 (Chapter 2 of this thesis)

Müller DG, Stache B, 1989: Life history studies on *Pilayella littoralis* (L.) Kjellman (Phaeophyceae, Ectocarpales) of different geographical origin.- Botanica Marina 32, 71-8

Muller P, Li XP, Niyogi KK, 2001: Non-photochemical quenching. A response to excess light energy.- Plant Physiology 125(4), 1558-66

Nedbal L, Soukupova J, Kaftan D, Whitmarsh J, Trtilek M, 2001: Kinetic imaging of chlorophyll fluorescence using modulated light.- Photosynthesis Research, *in press*

Noctor G, Foyer CH, 1998: Ascorbate and glutathione: keeping active oxygen under control.- Annual Review of Plant Physiology & Plant Molecular Biology 49, 249-79

Omasa K, Shimazaki K-I, Aiga I, Larcher V, Onoe M, 1987: Image analysis of chlorophyll fluorescence transients for diagnosing the photosynthetic system of attached leaves. – Plant Physiol. 84, 748-52

Oxborough K, Baker NR, 1997a: Resolving chlorophyll a fluorescence images of photosynthetic efficiency into photochemical and non-photochemical components-calculation of q_P and F_v'/F_M' without measuring F_0'.- Photosynth. Res. 54, 135-42

Oxborough K, Baker NR, 1997b: An instrument capable of imaging chlorophyll alpha fluorescence from intact leaves at very-low irradiance and at cellular and subcellular levels of organization. - Plant Cell Environ. 20(12), 1473-83

Pearson G, Kautsky L, Serrão E, 2000: Recent evolution in Baltic *Fucus vesiculosus*: reduced tolerance to emersion stresses compared to intertidal (North Sea) populations.- Marine Ecology Progress Series 202, 67-79

Peterson RB, Aylor DE, 1995: Chlorophyll fluorescence induction in leaves of *Phaseolus vulgaris* infected with bean rust (*Uromyces appendiculatus*).- Plant Physiol. 108, 163-71

Raggi V, 1995: CO_2 assimilation, respiration and chlorophyll fluorescence in peach leaves infected by *Taphrina deformans*.- Physiologia Plantarum 93, 540-4

Ravenel J, Peltier G, 1991: Inhibition of chlororespiration by myxothiazol and antimycin A in *Chlamydomonas reinhardtii*.- Photosynthesis Research 28, 141-8

Roberts AM, Walters DR, 1988: Photosynthesis in discrete regions of leek leaves infected with the rust, *Puccinia allii* Rud.- New Phytologist 110, 371-6

Robledo DR, Sosa PA, García-Reina G, Müller DG, 1994: Photosynthetic performance of healthy and virus-infected *Feldmannia irregularis* and *F. simplex* (Phaeophyceae).- Eur. J. Phycol. 29, 247-51

Rolfe SA, Scholes JD, 1995: Quantitative imaging of chlorophyll fluorescence.- New Phytologist 131, 69-79

Scholes JD, Farrar JF, 1986: Increased rates of photosynthesis in localised regions of a barley leaf infected with brown rust.- New Phytologist 104, 601-12

Scholes JD, Rolfe, SA, 1996: Photosynthesis in localised regions of oat leaves infected with crown rust (*Puccinia coronata*): quantitative imaging of chlorophyll fluorescence.- Planta 199, 573-82

Schreiber U, Neubauer C, Schliwa U, 1993: PAM fluorometer based on medium-frequency pulsed Xe-flash measuring light: A highly sensitive new tool in basic and applied photosynthesis research.- Photosynthesis Research 36, 65-72

Schreiber U, Hormann H, Neubauer C, Klughammer C, 1995: Assessment of photosystem II photochemical quantum yield by chlorophyll fluorescence quenching analysis.- Australian Journal of Plant Physiology 22, 209-20

Siebke K, Weis E, 1995: Assimilation images of leaves of *Glechoma hederacea*: Analysis of non-synchronous stomata related oscillations.- Planta 196, 155-65

Starr RC, Zeikus JA, 1987: UTEX - the culture collection of algae at the University of Texas at Austin.- J. Phycol. 23 (Suppl.), 1-47

van Kooten O, Snel JFH, 1990: The use of chlorophyll fluorescence nomenclature in plant stress physiology.- Photosynthesis Research 25, 147-50

Young AJ, Frank HA. Energy transfer reactions involving carotenoids: Quenching of chlorophyll fluorescence. Journal of Photochemistry and Photobiology B - Biology 36(1), 3-15

Yurina TP, Karavaev VA, Solntsev MK, Kukushkina MA, Ekobena FAP, 1996: Physiological features of the leaves of wheat cultivars resistant and susceptible to powdery mildew.- Fiziologiya Rastenii (Moscow) 43(1), 73-9

Figures

Fig. 1. Infections of *Eurychasma dicksonii* (Oomycota) and *Chytridium polysiphoniae* (Chytridiomycota) in *Pylaiella littoralis*, shown in images recorded with a digital video (CCD) camera (recorded at low irradiance, in order to keep the cells dark-adapted). **A)** *Young* stages of *Eurychasma dicksonii* (i.e. max. diameter of infected cells / diameter of uninfected cells = 1.44 ± 0.25). **B)** *Medium* stages of *Eurychasma dicksonii* (i.e. max. diameter of infected cells / diameter of uninfected cells = 2.07 ± 0.44). **C)** *Late* stages of *Eurychasma dicksonii* (i.e. max. diameter of infected cells / diameter of uninfected cells = 2.80 ± 0.45). **D)** *Chytridium* infections, about one week after spore settlement.-
Scale bars correspond to 25 μm.

Fig. 2. Change of fluorescence parameters in *Eurychasma*-infected cells of *Pylaiella littoralis*. First column: uninfected cells; second column: cells infected by young *Eurychasma* stages; third column: cells infected by medium *Eurychasma* stages; fourth column: cells infected by old *Eurychasma* stages. F_V / F_M is a parameter for the maximum photochemical efficiency of PS II (Björkman & Demmig, 1987), (F_M – F_S) / F_M represents a vitality index (Lichtenthaler & Rinderle, 1988), whilst F_0 / F_0' is an indicator for non-photochemical quenching (Muller *et al.*, 2001).

Fig. 3. Change of fluorescence parameters in *Chytridium*-infected cells of *Pylaiella littoralis*. First column: uninfected cells; second column: cells infected by young stages of *Chytridium*. With further development of the fungal sporangia, photosynthetic activity in the parasitized *Pylaiella* cells ceased completely, with a total loss of photosynthetic pigments.

Fig. 4. Light saturation (PI) curve of an *Eurychasma*-infected (dashed line, squares) and a neighbouring, uninfected (continuous line, diamonds) *Pylaiella* cell, representative for one out of six experiments. F_P / F_0 indicates the light-saturation curve of PS II photochemistry. The consistently lower values for *Eurychasma*-infected cells observed in these experiments show an increased dark adaptation.

Fig. 5. Comparison of Kautsky curves (***up***) and their derivatives (***below***). The symbols in the upper graphs represent the individual measuring points, while the lines (S20AA1, S20AA2) represent the corresponding smoothed curves,. For which the data were treated by twice using the adjacent averaging (AA) method with 20 points.

A) Changes of Kautsky slopes as a function of infection stage in the case of *Eurychasma*-infected *Pylaiella*. ***Left:*** early infection + control; ***middle:*** medium infection stage + control; ***right:*** late infection stage + control

B) Changes of Kautsky slopes as a function of irradiance in the case of *Eurychasma*-infected *Pylaiella*. ***Left:*** low (non-saturating) irradiance; ***middle:*** saturating irradiance; ***right:*** over-saturating (photoinhibitory) irradiance

C) Changes of Kautsky slopes as a function of infection stage in the case of *Chytridium*-infected *Pylaiella*. ***Left:*** early infection + control; ***middle:*** medium infection stage + control; ***right:*** late infection stage + control

Fig. 6. Illustration of the FKM technique, using the example of an *Eurychasma*-infected *Pylaiella* filament. **A)** Colour image taken with digital photo camera in transmittant light. **B)** Greyscale image taken with digital film (CCD) camera in a weak, non-actinic transmittant light. **C)** False colour image of chlorophyll fluorescence calculated from fluorescence kinetic film. **D)** Manual selection of objects for kinetic analysis using the FluorCam™ software. In the chosen mode, all pixels within the selected areas will be averaged. **E)** Fluorescence induction (Kautsky) curves of objects 1, 2, 4, 6 calculated by FluorCam™ software clearly showing the differences between infected and uninfected cells, with distinctly different curvatures.

Figure 1

A

B

C

D

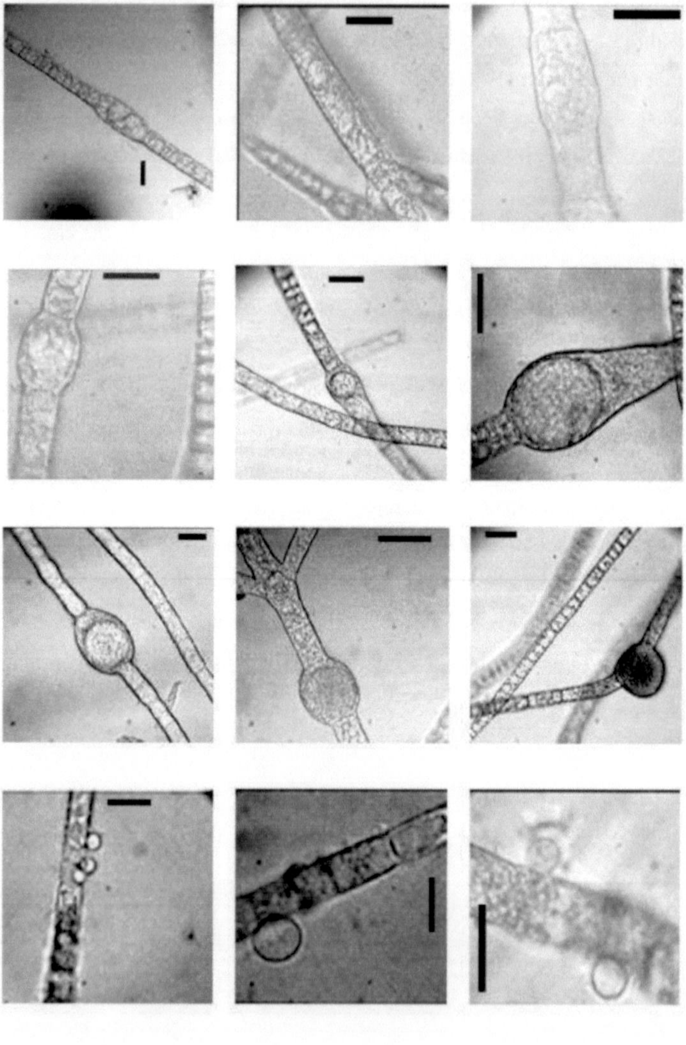

Figure 2

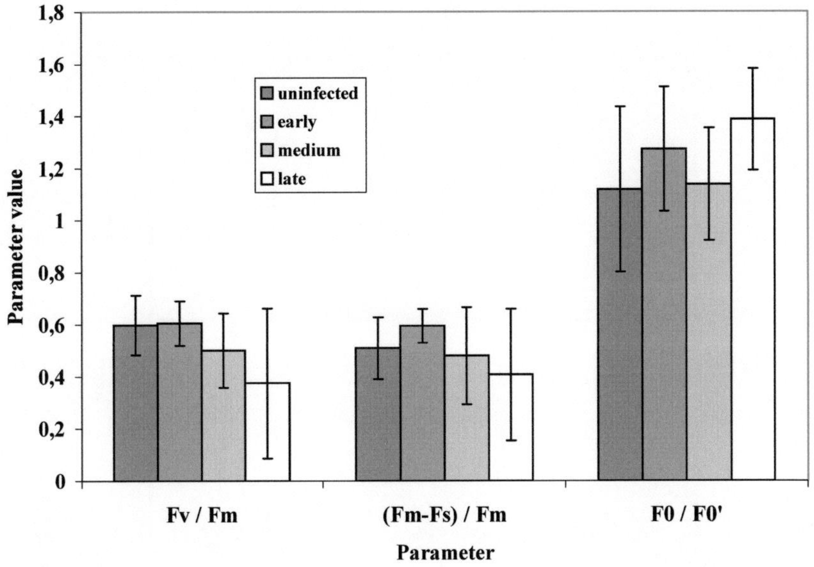

Figure 3

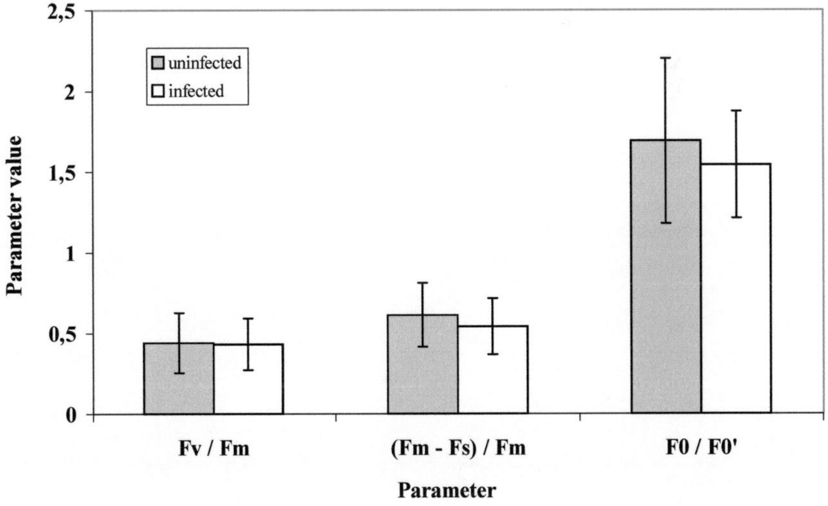

Figure 4

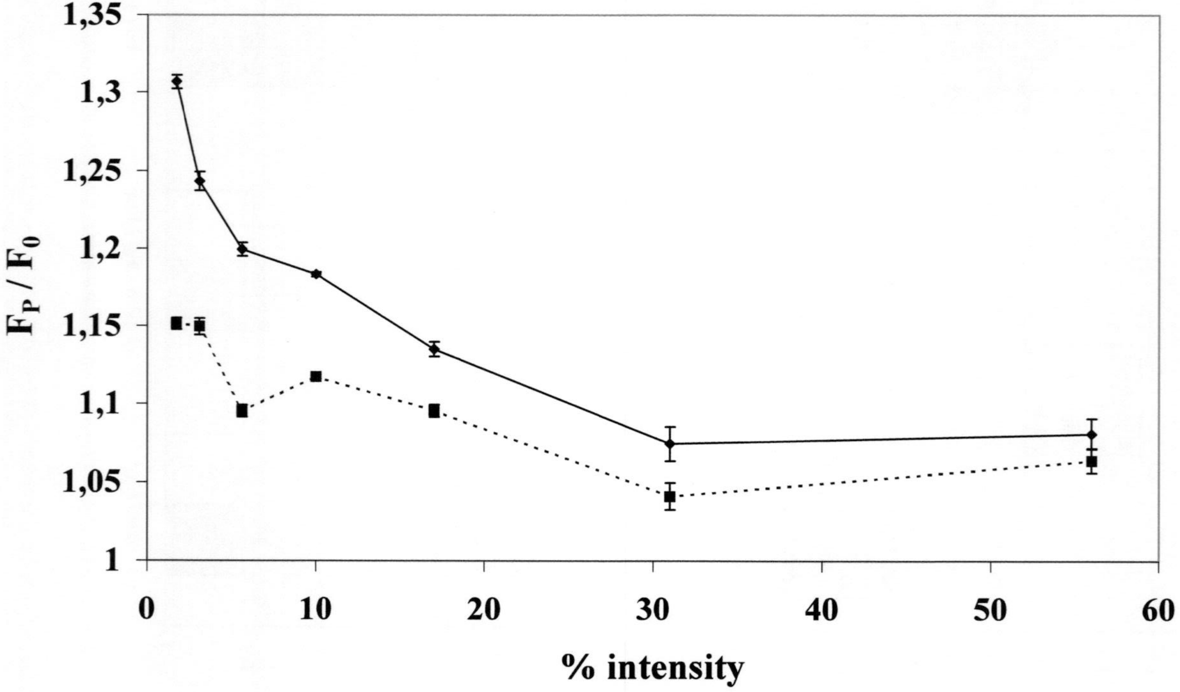

Figure 5 A

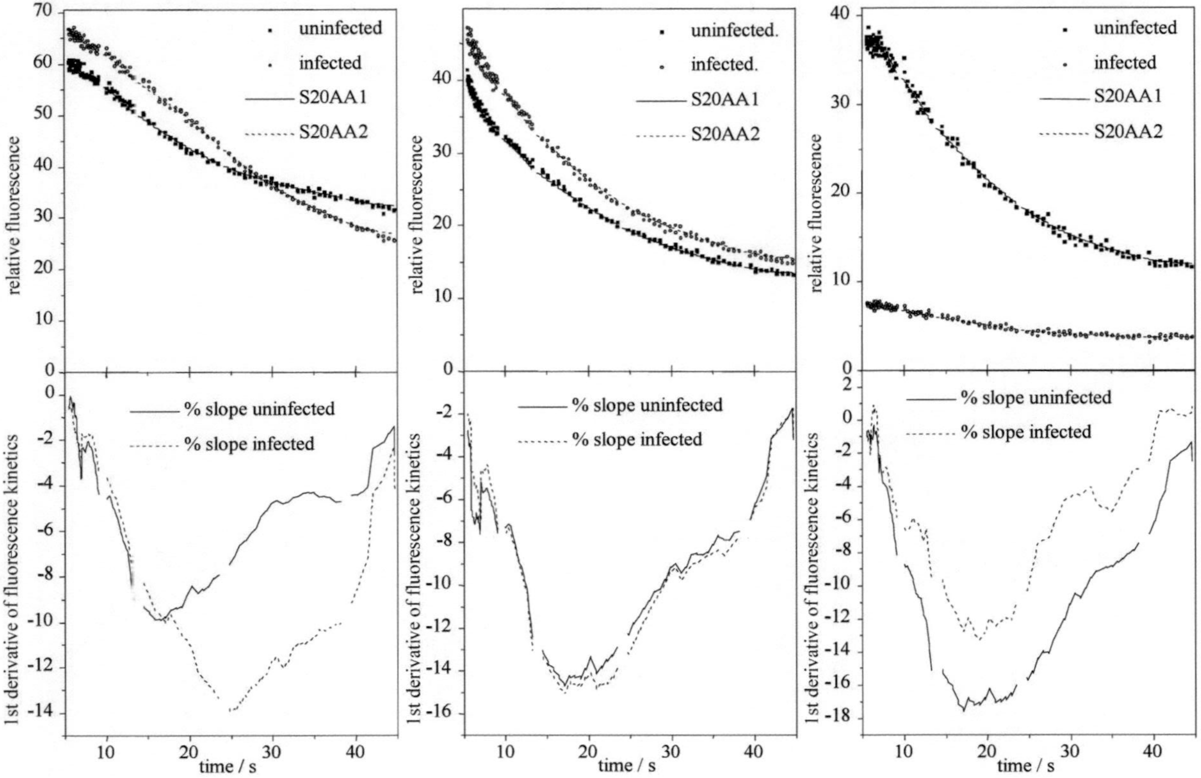
uninfected
infected
S20AA1
S20AA2
uninfected.
infected.
% slope uninfected
% slope infected
relative fluorescence
1st derivative of fluorescence kinetics
time / s

Figure 5 B

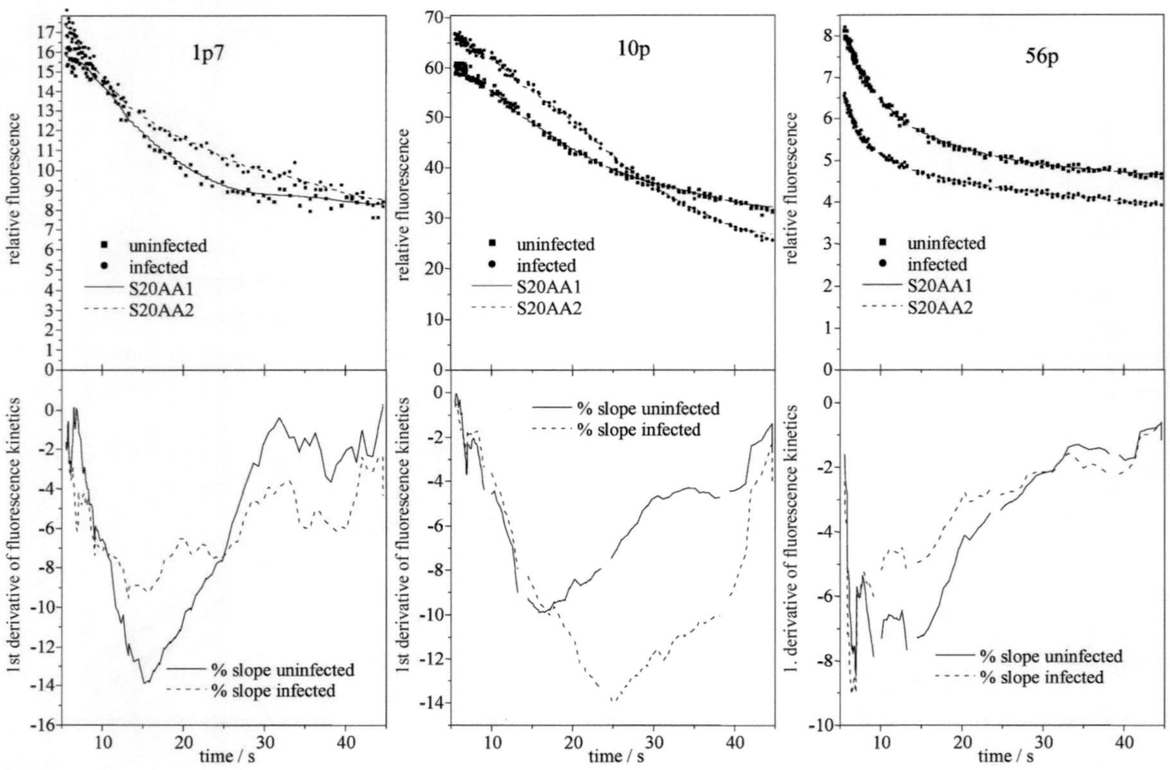
1p7
10p
56p
relative fluorescence
uninfected
infected
S20AA1
S20AA2
1st derivative of fluorescence kinetics
1. derivative of fluorescence kinetics
% slope uninfected
% slope infected
time / s

Figure 5 C

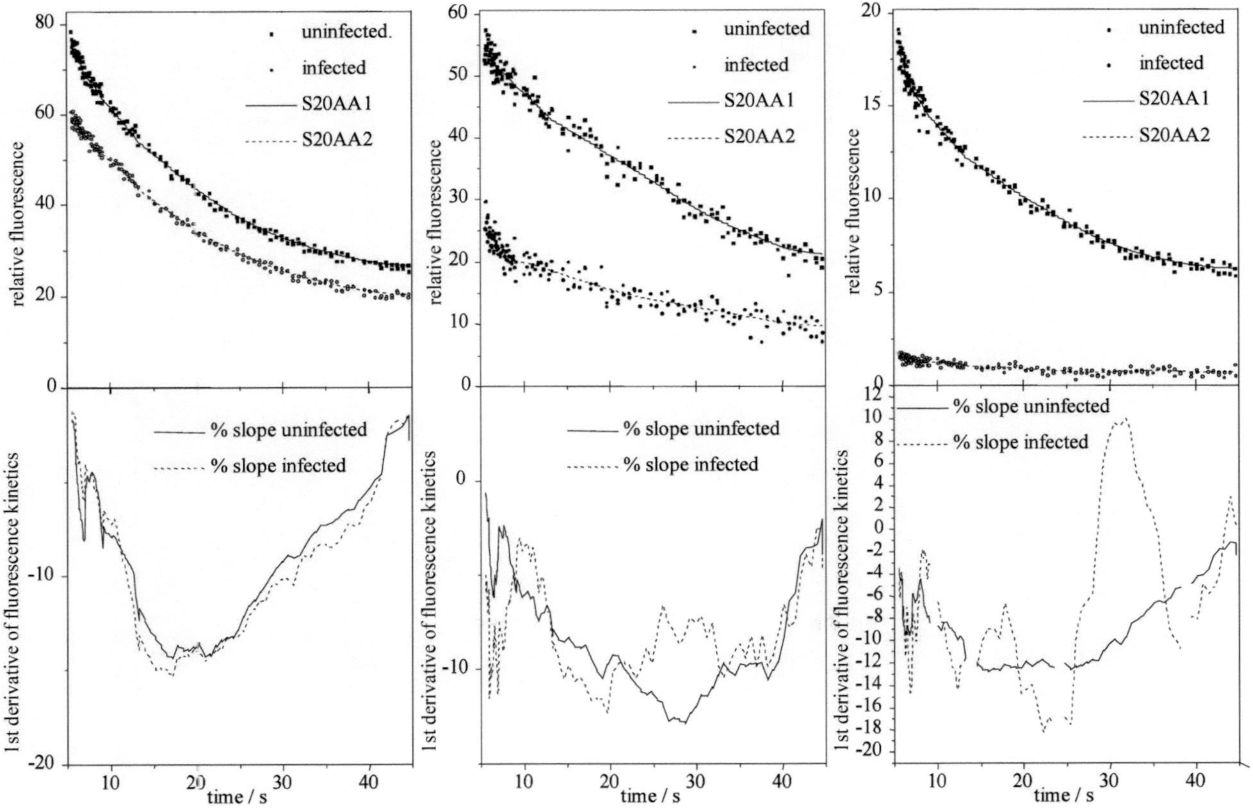
uninfected.
infected
S20AA1
S20AA2
relative fluorescence
% slope uninfected
% slope infected
1st derivative of fluorescence kinetics
time / s
uninfected
infected
S20AA1
S20AA2
relative fluorescence
% slope uninfected
% slope infected
1st derivative of fluorescence kinetics
time / s
uninfected
infected
S20AA1
S20AA2
relative fluorescence
% slope uninfected
% slope infected
1st derivative of fluorescence kinetics
time / s

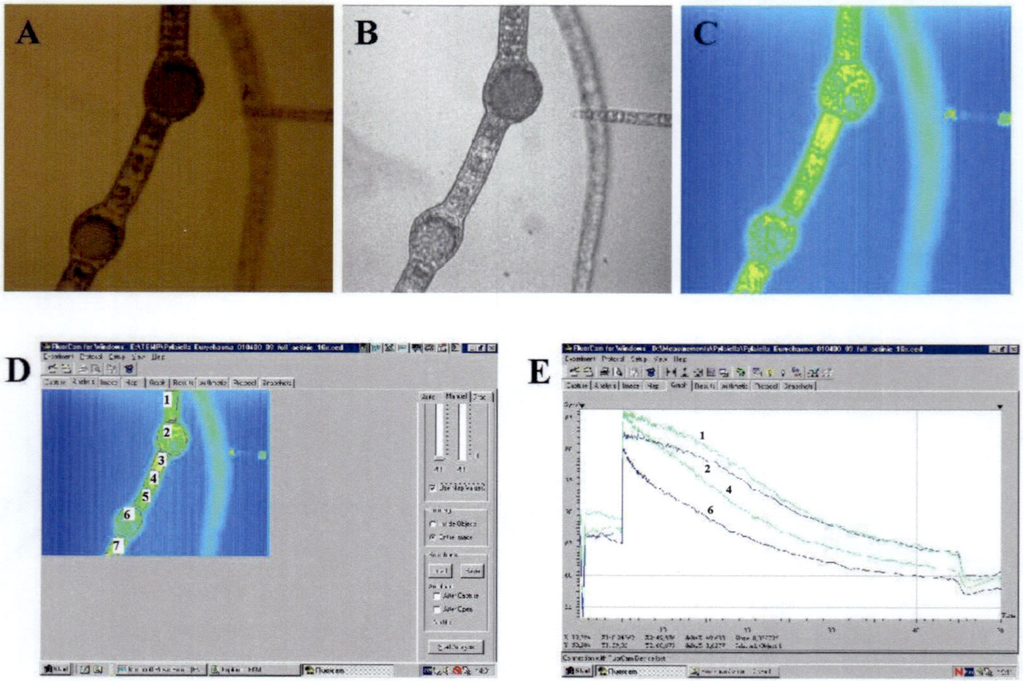

Figure 6

6

Oligoguluronates Elicit an Oxidative Burst in the Brown Algal Kelp *Laminaria digitata*[1]

Frithjof Christian Küpper, Bernard Kloareg, Jean Guern, and Philippe Potin*

Unité Mixte de Recherche, 1931 du Centre National de la Recherche Scientifique, and Laboratoires Goëmar, Station Biologique, Place G. Teissier, BP 74, F–29682 Roscoff, Brittany, France (F.C.K., B.K., P.P.); and Institut des Sciences Végétales, Centre National de la Recherche Scientifique, avenue de la Terrasse, F–91198 Gif-sur-Yvette, France (J.G.)

Oligomeric degradation products of alginate elicited a respiratory and oxidative burst in the sporophytes of the kelp *Laminaria digitata*. The generation of activated oxygen species (AOS), O_2^-, and H_2O_2 was detected at the single cell level, using nitroblue tetrazolium precipitation and a redox-sensitive fluorescent probe, respectively. The oxidative burst involved diphenyleneiodonium-sensitive AOS-generating machinery and its amplitude depended on the type of tissue. After a first elicitation plants were desensitized for about 3 h. The activity of alginate oligosaccharides was dose dependent, saturating around 40 μM. It was also structure-dependent, with homopolymeric blocks of α-1,4-L-guluronic acid, i.e. the functional analogs of oligogalacturonic blocks in pectins, being the most active signals. The perception of oligoguluronate signals resulted in a strong efflux of potassium. Pharmacological dissection of the early events preceding the emission of AOS indicated that the transduction chain of oligoguluronate signals in *L. digitata* is likely to feature protein kinases, phospholipase A_2, as well as K^+, Ca^{2+}, and anion channels.

The oxidative burst, i.e. a rapid, transient production of large amounts of activated oxygen species (AOS) such as superoxide (O_2^-), hydrogen peroxide (H_2O_2) or hydroxyl radicals (OH) was discovered several decades ago as a "respiratory burst" during phagocytosis in cells of the human immune system (Baldridge and Gerard, 1933). The same phenomenon was found later in plants (Doke, 1983a, 1983b) and it is recognized today as a ubiquitous characteristic of defense systems in these phyla (e.g. Baker and Orlandi, 1995; Bolwell et al., 1995; Lamb and Dixon, 1997; Wojtaszek, 1997). This defense response is most often triggered by cell-cell recognition, involving the perception at the plant cell membrane of signal molecules from the invading organism or from the host cell walls, referred to as elicitors. Common elicitors of non-host resistance are oligosaccharides, glycoproteins, and glycopeptides (e.g. Scheel and Parker, 1990; Boller, 1995). In higher plant pathogen recognition systems, xyloglucan (Fry et al., 1993) and pectin (e.g. Boudart et al., 1998) elicitors are of endogenous origin (i.e. from the plant), whereas glucans and chitins are exogenous (released from the pathogen during the interaction).

An increasing amount of scientific evidence (for review, see Wojtaszek, 1997) points out that the onset of the oxidative burst is controlled by a signal transduction cascade involving G proteins, adenylate cyclases, phospholipases, protein kinases, protein phosphatases, ion channels, changes in membrane potential, permeability and ion fluxes, and finally the activation of AOS-generating oxidases/dehydrogenases. The main enzymatic source of AOS in the oxidative burst of plant cells remains controversial, but most results point to three mechanisms: (a) an O_2^--generating plasmalemmic NAD(P) H oxidase (Desikan et al., 1996; Groom et al., 1996; Pugin et al., 1997; Jabs et al., 1997; Keller et al., 1998; Torres et al., 1998), analogous to that of mammalian cells (Morel et al., 1991); (b) other, apoplastic oxidases, such as oxalate oxidase (e.g. Dumas et al., 1993; Zhang et al., 1995; Thordal-Christensen et al., 1997) and amine oxidase (Allan and Fluhr, 1997); and (c) pH-dependent apoplastic peroxidases (Bolwell et al., 1995; 1998; Frahry and Schopfer, 1998; Martinez et al., 1998), which generate either O_2^- anions or hydrogen peroxide.

The high concentrations of AOS generated in the oxidative burst have direct, cytotoxic effects on invading pathogens (Peng and Kùc, 1992). In addition, the oxidative burst is known to orchestrate a variety of secondary defense responses, which include the production of low-M_r compounds with antimicrobial activity referred to as phytoalexins (Ebel et al., 1995; Apostol et al., 1987), synthesis of pathogenesis-related proteins (Hammond-Kossack and Jones, 1996; Fritig et al., 1998), and crosslinking of cell wall proteins (Brisson et al., 1994; Otte and Barz, 1996). Oxy-

[1] This work was supported by the Fonds Européens de Développement Régional (agricultural region; Vb) program of the European Commission. F.C.K. was generously supported with fellowships by Studienstiftung des Deutschen Volkes (Bonn), Hüls AG-Stiftung (Marl, Germany), and the European Commission (Program MAST-III, Brussels).

* Corresponding author; e-mail potin@sb-roscoff.fr; fax 33–2–98–29–23–24.

278 *Plant Physiology*, January 2001, Vol. 125, pp. 278–291, www.plantphysiol.org © 2001 American Society of Plant Physiologists

gen radicals are also thought to be involved in the hypersensitive response, i.e. limited necrosis at the site of pathogen attack (Levine et al., 1994; Tenhaken et al., 1995; Richberg et al., 1998), as well as in systemic acquired resistance (Chen et al., 1993; Dempsey and Klessig, 1994; Van Camp et al., 1998).

In contrast there is only limited knowledge about the cell-cell recognition mechanisms and chemical defenses involved in host-pathogen interactions in the marine environment (Potin et al., 1999) and we do not know whether the concepts derived from the study of terrestrial plant pathosystems also apply to the distantly related lineages found in the sea. A few observations, however, indicate that oligosaccharide recognition and the oxidative burst mediate host-microbe interactions in marine algae, too. Oxidative burst per se was first reported in the red alga *Eucheuma platycladum* following mechanical injury (Collén and Pedersén, 1994). More recently, oxidative burst induced by oligosaccharide elicitors was shown to induce protection against parasites in the red algae *Chondrus crispus* (Bouarab et al., 1999) and *Gracilaria conferta* (Weinberger et al., 1999), and AOS were demonstrated to play a key role in programmed cell death in blooms of the phytoplankton dinoflagellate *Peridinium gatunense* (Vardi et al., 1999).

This study was undertaken to investigate whether such phenomena also exist in brown algae, using as model system *Laminaria digitata*, a common kelp on the rocky shores of the North Atlantic. Oligosaccharides derived from alginate, the main brown algal cell wall polysaccharide (Fig. 1), are shown to elicit a marked oxidative burst in the cortical cells of *L. digitata* sporophytes, sufficient to control populations of epiphytic bacteria. Pharmacological evidence indicates that the transduction of alginate oligosaccharide signals involves some of the steps described for higher plants (Wojtaszek, 1997; Scheel, 1998), leading to the activation of a diphenylene iodonium chloride (DPI)-sensitive AOS-generating system.

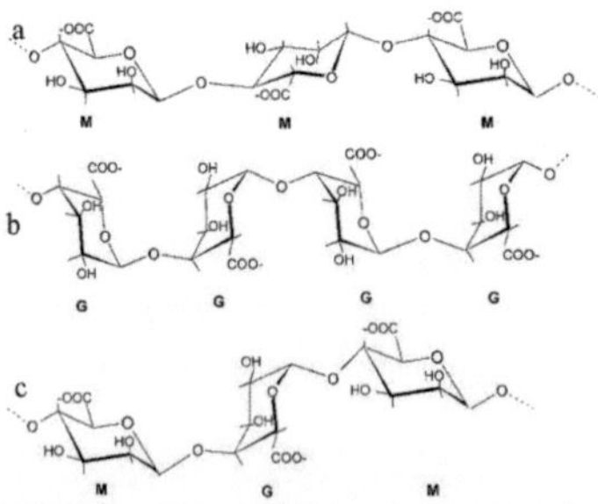

Figure 1. Chemical structure of alginates from brown algae. Alginates are linear anionic copolymers of β-1,4-D-mannuronic acid and of its C_5 epimer, α-1,4-L-guluronic acid. They consist of the alternation of homopolymeric blocks of poly-β-1,4-D-mannuronic acid, referred to here as MM blocks (a), of homopolymeric blocks of poly-α-1,4-L-guluronic acid (GG blocks; b), and of heteropolymeric blocks with random arrangements of both monomers (MG blocks; c).

RESULTS

Oligoguluronates Induce a Strong Respiratory and Oxidative Burst in *L. digitata* Sporophytes

Treatment of young sporophytic thalli of *L. digitata* with alginate oligosaccharides induced a massive increase in oxygen consumption within 2 to 3 min of their addition (Fig. 2A). As shown in Figure 2A, the increase in oxygen consumption was sometimes bimodal, first a sharp respiratory burst, with an initial rate up to 8 times that of basal respiration, followed a few minutes later by a second, less intense increase of apparent respiration. In the presence of oligoguluronates at the concentration of 50 μg mL^{-1}, the increase in oxygen consumption typically lasted 10 min, amounting to a total extra consumption of approximately 5 μmol O_2 g^{-1} fresh weight. Only the guluronate-containing alginate oligosaccharides, i.e. GG blocks and MG blocks, were able to elicit a respiratory response in *L. digitata* sporophytes. In contrast, MM blocks transiently decreased the apparent respiration (Fig. 2A).

At the same time a sudden and strong release of H_2O_2 into the incubation medium was observed (Fig. 2B). The peroxide concentration in the medium reached its maximum 5 to 10 min after the addition of alginate fragments, then decreased progressively back to the initial level after about 40 to 50 min. Addition of catalase (100 U mL^{-1}) prior to GG fragments completely inhibited the signal in the luminol assay (data not shown), indicating that this assay primarily detected H_2O_2. Superoxide dismutase (SOD; 200 U mL^{-1}) did not have a significant effect on the signal. The oxidative response was again dependent on the structure of the alginate oligosaccharides used as signal. MM did not elicit H_2O_2 release in *L. digitata* sporophytes (Fig. 2B). In contrast, when applied at the concentration of 50 μg mL^{-1} in experiments involving a biomass ratio of 20 to 50 mg fresh weight mL^{-1}, oligoguluronates typically initiated release of H_2O_2 up to concentrations in the 10 μM range. MG were also active in eliciting an oxidative response, yet to a much lesser extent than oligoguluronates (Fig. 2B).

The initial rate and amplitude of H_2O_2 release were also dependent on the concentration of alginate oligosaccharides. The threshold of elicitor concentration required to trigger an oxidative burst was between 1 (not yet active) and 2.5 μg mL^{-1} (already triggering a response), whereas the response saturated at about 150 μg mL^{-1} (Fig. 2C). A desensitizing effect was observed after the first elicitation: after a first elicita-

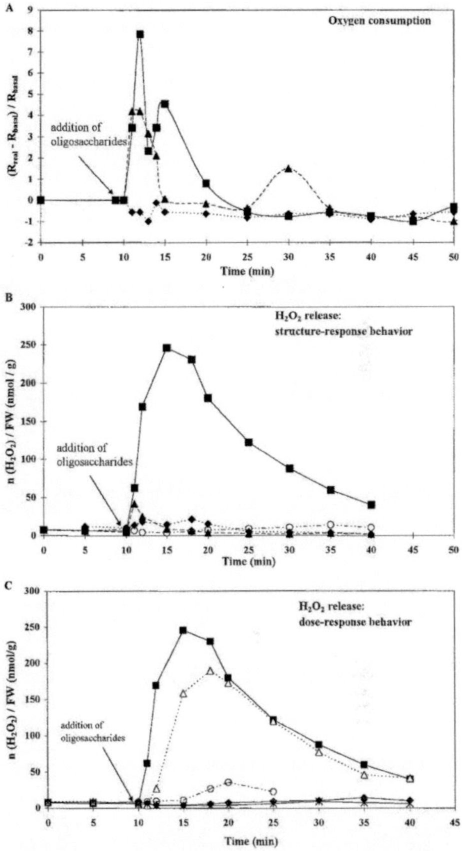

Figure 2. Oligoalginate-induced respiratory and oxidative bursts in *L. digitata*. A, Oligoguluronates stimulate oxygen consumption in young *L. digitata* plants. Young *L. digitata* plantlets were elicited with 50-μg mL^{-1} G blocks (■), MG (▲) blocks, or M blocks (◆), respectively. Their oxygen consumption was recorded with a Clark-type oxygen electrode and expressed relative to basal respiration (with R_{real} as the respiration rate at a given moment and R_{basal} as the basal respiration rate averaged over 15 min prior to elicitation). B, Oxidative burst in *L. digitata*: structure-response relationships. Kinetics of H_2O_2 release of *L. digitata* fronds challenged with G blocks (■), MG blocks (▲), M blocks (◆), and seawater (control, ○). C, Oxidative burst in *L. digitata*: dose-response behavior. Hydrogen peroxide production by *L. digitata* plantlets, after elicitation with 1 μg mL^{-1} (crosses), 10 μg mL^{-1} (△), and 100 μg mL^{-1} (■), respectively (control, ◆). The threshold for triggering a burst was at 2.5 μg mL^{-1} (○).

tion producing a burst, subsequent challenges with alginate oligosaccharides remained ineffective over a period of at least 3 h. When transferred to running seawater, elicited plantlets did not develop visible necroses in culture.

DPI, an irreversible inhibitor of the mammalian neutrophil NADPH oxidases (O'Donnell et al., 1993), inhibited the oxidative response in *L. digitata* sporophytes when used at a concentration of 10 μM (Fig. 3). Quinacrine, a general inhibitor of flavoprotein

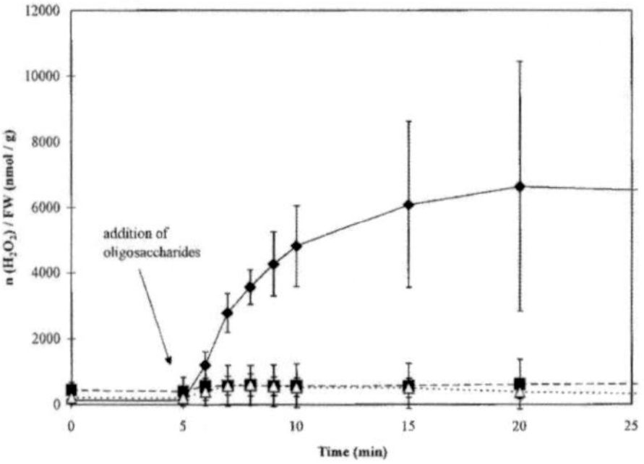

Figure 3. Effect of DPI on the oxidative burst in *L. digitata*. Kinetics of H_2O_2 emission by *L. digitata* plantlets challenged with 150 μg mL^{-1} GG (◆), 150 μg mL^{-1} GG after pre-incubation for 15 min with 10 μM DPI (△), and by unchallenged plantlets (■).

oxidases (Auh and Murphy, 1995), also blocked H_2O_2 release (data not shown).

Histology of the Oxidative Burst in *L. digitata* Sporophytes

Based on analyses by confocal microscopy with the fluorescent, redox-sensitive dye dichlorohydrofluorescin, AOS formation was triggered as soon as 1 min after elicitation of *L. digitata* plantlets with oligoguluronates and increased exponentially for at least 7 min. AOS mainly accumulated around epidermal and outer cortical cells, but not in the medulla (Fig. 4). Staining *L. digitata* with nitroblue tetrazolium (NBT) in the presence of oligoguluronates resulted in a blue precipitate, indicating that superoxide anions are formed during the oxidative burst (Fig. 5A). As indicated by a deeper blue stain, the accumulation of superoxide was enhanced by diethyldithiocarbamate (DDC), a strong inhibitor of SOD (Auh and Murphy, 1995; Jabs et al., 1997; Fig. 5C). However, attempts to prevent NBT accumulation by the addition of exogenous SOD were unsuccessful. In contrast and consistent with the effect of DPI on the release of hydrogen peroxide, pretreatment of young plantlets with this inhibitor (10 μM DPI, 15 min) did decrease precipitation of NBT upon elicitation with GG blocks (Fig. 5E). In agreement with the results of confocal microscopy, the staining with NBT of cross sections of *L. digitata* plantlets showed that O_2^- was accumulated mainly in the cortical cell layers, whereas it was hardly detectable in the medulla (Fig. 5, F and G).

Strong tissue-specific differences were observed when various tissues of adult *L. digitata* sporophytes were compared in their response with oligoguluronates. The young blade tissues were the most sensitive and the meristematic and older blade tissues were the least reactive parts. The sensitivity appeared to be confined to the cortex, the medulla (i.e. stipe tissue without cortex) showing no response upon elicitation (Fig. 6). Only *L. digitata* sporophytes and sporophyte suspended-cell cultures (data not shown) reacted with an oxidative burst upon elicitation with alginate oligosaccharides. *L. digitata* gametophytes, the filamentous, haploid life history phase, did notrespond to the presence of oligoalginate signals (data not shown).

Pharmacological Analysis of the Oxidative Burst in *L. digitata* Sporophytes

In an attempt to unravel the signal transduction pathways leading to AOS formation upon elicitation with alginate oligosaccharides in *L. digitata*, a number of compounds known to affect these processes in higher plants were tested. Pre-incubation of *L. digitata* plantlets in the presence of 10 μM staurosporine, a general inhibitor of protein kinases (Tavernier et al., 1995), inhibited the emission of H_2O_2 following elicitation with oligoguluronates. Inhibitors of calcium channels such as verapamil and methoxyverapamil effectively blocked the induction of the oxidative burst. The Ca^{2+} ionophore A23187 induced a strong oxidative burst, about 3 times higher than the control response to GG blocks. The inhibitor of higher plant anion channels, A9C (Cazalé et al., 1998), inhibited the burst, and so did two other anion channel inhibitors, 4,4′-diisothiocyanatestilbene-2, 2′-disulfonic acid (DIDS) and 5-nitro-2-(3-phenylpropylamino-) benzoic acid (NPPB; Lurin et al., 1996; Jabs et al., 1997). Chlorpromazine-HCl, an antagonist of phospholipase A_2 in higher plants (Chandra et al., 1996), completely blocked the induction of the response in *L. digitata*. Mastoparan, a specific activator of G proteins (Legendre et al., 1992, 1993), failed to

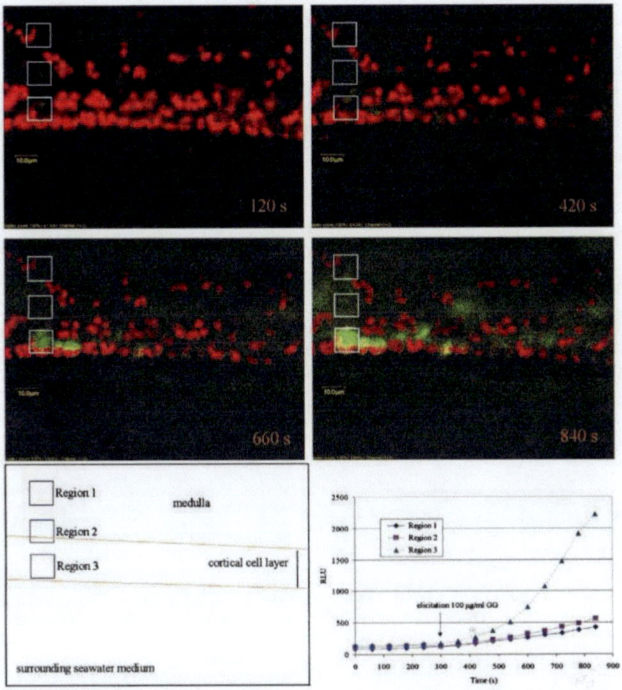

Figure 4. Laser scanning confocal imaging of elicited oxidative burst in *L. digitata* plantlets. *L. digitata* plantlets were loaded with the redox-sensitive fluorescent probe dichlorohydrofluorescein diacetate (DCFH-DA, Molecular Probes, Eugene, OR) and examined by laser scanning confocal microscopy. Elicitors were added at $t = 300$ s during the time course of image acquisition. The green fluorescence of dichlorofluorescein was monitored on channel 1, concomitant with the red fluorescence of plastids on channel 2. Four images are presented, representative of the time course of fluorescence with flurorescences in channels 1 and 2 overlaid. Squares labeled 1, 2, and 3, respectively, were delineated from the outer cortex to the medulla, to record the three curves showing the respective integrated fluorescence. The data reported here correspond to one representative experiment out of three (bar corresponds to 10 μm and RLU refers to relative light units defined by the integrated luminosity of the pixels of the given area).

induce release of H_2O_2 from *L. digitata* sporophytes, even at concentrations as high as 10 μM.

All of the specific ionophores tested, namely valinomycin and nonactin (K^+), monensin (Na^+), A23187 ionophore (Ca^{2+}), and dinitrophenol (H^+) enhanced the oxidative response of *L. digitata* to oligoguluronate signals (Table I). When used alone at the concentration of 57 μM, valinomycin induced a strong oxidative burst, about one order of magnitude stronger than the usual response to GG blocks (Fig. 7). Both signals were capa-

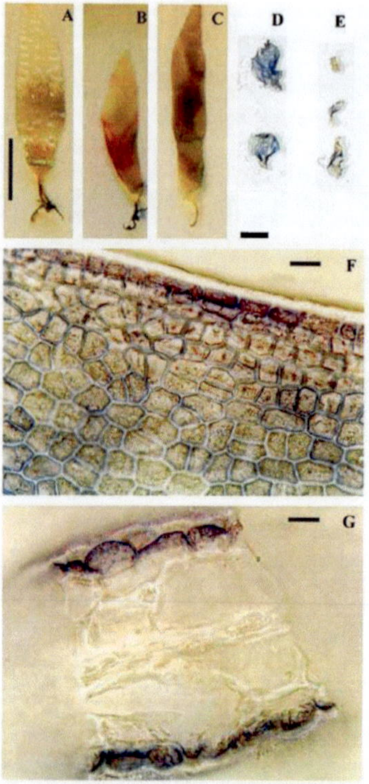

Figure 5. Superoxide generation in cortical tissues of *L. digitata* young sporophytes. After the addition of oligoguluronates, plantlets were stained with NBT from 15 to 30 min after elicitation to assay superoxide detected by a dark blue precipitate after bleaching of the plant pigments. The plants were treated as follows: A (control), pre-incubation with DDC, but without previous elicitation (almost no NBT staining due to low steady-state superoxide production); B, elicitation with 50 μg mL^{-1} GG (NBT staining 30 min later; some O_2^- detectable, but most of it is rapidly converted to H_2O_2); C, inhibition of SOD by pre-incubation with 1 mM DDC and elicitation with 50 μg mL^{-1} GG (NBT staining 15 min later; intensive stain due to massive accumulation of O_2^-); D, elicitation with 50 μg mL^{-1} GG (NBT staining 15 min later) on younger cultivated plantlets (control assay for the panel on the right); E, pre-incubation with 10 μM DPI and elicitation with 50 μg mL^{-1} GG (NBT staining 15 min later) on younger cultivated plantlets (almost no accumulation of O_2^- because the enzymatic source is inhibited). In A through E, the bar corresponds to 1 cm. F, Superficial view of the cortical cells near the margin of the blade of a young *L. digitata* plantlet pre-incubated with DDC and stained with NBT 15 min after elicitation (bar corresponds to 20 μm). A strong blue precipitate is visible in the surroundings of the epidermal cells. G, Cross section of the same blade (bar corresponds to 25 μm), showing the accumulation of blue precipitate around the cortical cells.

ble of acting synergistically, triggering bursts two orders of magnitude higher than the oxidative response induced by oligoguluronates (Fig. 7).

Oligoguluronates Trigger Potassium Efflux in *L. digitata* Sporophytes

A marked potassium efflux was observed from *L. digitata* sporophytes challenged with oligoguluronates in artificial seawater partially depleted in potassium (Fig. 8). After a slight initial decrease of the K^+ concentration in the surrounding medium, a steady net K^+ efflux was established, corresponding to a loss of about 10 μmol K^+ per gram fresh weight within 1 h. Concomitantly, a slight alkalinization (0.07 pH units) was detectable in unbuffered media within the first 3 min after elicitation, followed by an external acidification continuing for at least 1 h, from pH 8.25 to 7.85 (Fig. 8). Overnight, the leakage of potassium from a 50 mg mL^{-1} biomass of *L. digitata* plantlets raised the external medium concentration by an average of 4.16 mM in elicited samples, compared with 1.67 mM in the controls.

H_2O_2 Generated by the Oxidative Burst Controls Growth of Kelp Pathogenic and Epiphytic Bacteria

The question of whether the amount of AOS generated by the oxidative burst in *L. digitata* can be toxic for pathogens was then indirectly tested by cultivation of an alginate-degrading bacterial strain associated with *Laminaria japonica*, *Pseudoalteromonas elyakovii* (Sawabe et al., 2000), and the epiphytic strain referred to as Ldm2 in the presence of various concentrations of H_2O_2. Both strains were completely inhibited in their growth by H_2O_2 concentrations as low as 10 μM. An H_2O_2 concentration of 250 μM was sufficient to kill a culture of *P. laminariocolax* and to reduce the survival of strain Ldm2 by 60%.

DISCUSSION

The Brown Algal Kelp *L. digitata* Recognizes Guluronate-Rich Alginate Fragments as Signals to Trigger an Oxidative Burst

Alginate (Fig. 1) is the main cell wall component of Laminariales, amounting to approximately 60% of *L. digitata* isolated cell walls (Mabeau and Kloareg, 1987). It consists of three types of blocks (Haug et al.,

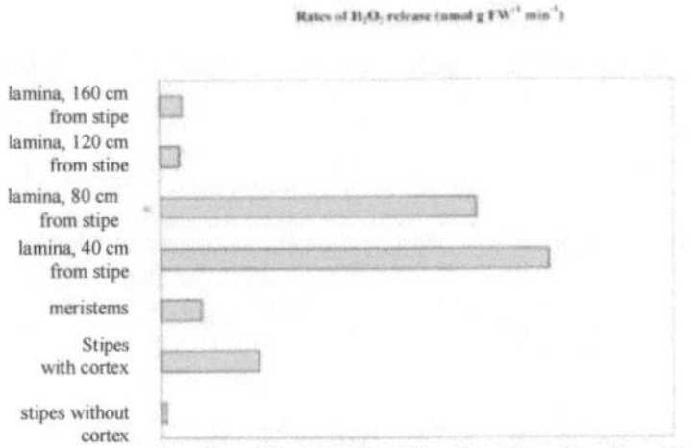

Figure 6. Comparison of the oxidative burst in various tissues of *L. digitata* after elicitation with oligoguluronates. Small tissue discs of an *L. digitata* plant were compared in their peak response with elicitation by GG blocks (50 μg mL^{-1}). The data reported here correspond to one representative experiment out of three.

1974), MM or GG, and blocks with mixtures of both monomers (MG). The above results demonstrate that reminiscent of the behavior of higher plants pectin oligosaccharides (e.g. Legendre et al., 1993), the sporophytes of the brown algal kelp *L. digitata* respond to incubation in the presence of alginate oligosaccharides by a sudden increase in oxygen consumption, concomitant with a marked release of hydrogen peroxide. The oxidative burst is rapid, starting within 2 to 3 min of the addition of oligosaccharides, and transient, lasting for no longer than 30 min.

External H_2O_2 concentrations reach their maximum 10 min after elicitation, i.e. at the same time or shortly after the peaks of oxygen consumption (Fig. 2, A and B). The average molar ratio between total oxygen consumption and the release of hydrogen peroxide upon the elicitation of *L. digitata* plantlets with alginate oligosaccharides, approximately five oxygen molecules consumed for the net release of one molecule of hydrogen peroxide, indicates that this brown alga rapidly detoxifies AOS during the oxidative burst. In higher plants detoxification of AOS involves protecting mechanisms such as the ascorbate-glutathione cycle, as well as scavenging enzymes such as catalases and peroxidases (Wojtaszek, 1997). No such mechanism has been described in brown algae yet. We are currently exploring the possibility that the mobilization of the intracellular iodine reservoir (Küpper et al., 1998) participates in the protection against oxidative stress in Laminariales (A. Baker, L. Carpenter, B. Kloareg, F. Küpper, P. Liss, and G. Malin, unpublished data).

L. digitata sporophytes react to the presence of alginate oligosaccharides in a dose-response dependent manner. Given that the oligoguluronate blocks used in these experiments had an average degree of polymerization of 20 to 25 (Heyraud et al., 1996), the threshold signal concentration required to elicit a significant burst is 0.5 μM, the oxidative response typically saturating around 40 μM (Fig. 2C). A clear structure-activity relationship was observed in the capacity of alginate oligosaccharides to elicit an oxidative burst in *L. digitata*. Alginate polymers and MM elicited no response. Only the homo-oligomeric fragments composed of GG and their co-oligomers with MG were recognized, with GG eliciting a stronger response than MG (Fig. 2, A and B). These observations indicate that stretches of guluronate residues are essential in the recognition of alginate oligosaccharides by *L. digitata*.

Oligoguluronate fractions obtained with the help of an alginate lyase, i.e. with unsaturated uronic acid residues at their non-reducing ends, were as active as those prepared by mild acid hydrolysis, indicating that the configuration of the non-reducing ends is not involved with signal recognition. However, the biodegradation products of sulfated fucans, a cell wall matricial polysaccharide that can be thought of as a structural and a functional analog of the fucoxyloglucans of higher plants (Kloareg and Quatrano, 1988), did not elicit any oxidative response in *L. digitata* sporophytes. Laminaran, the low molecular weight β-1,3-glucan storage polysaccharide of brown algae, which structurally mimics β-1,3–1,6-glucan fungal elicitors is known to elicit a variety of defense responses in higher plants (*Ebel, 1998*). Based on H_2O_2 assays, laminaran was not recognized as a defense signal by *L. digitata* sporophytes (data not shown). Altogether, alginate fragments made of guluronate residues appear as a relatively specific defense signal in *L. digitata*. It is worth noting here that the ordered conformation of poly-α-1,4-L-guluronic acid is very similar to that of the poly-α-1,4-L-GalUA blocks of pectins, both leading to intermolecular associations in the presence of calcium ions (Kohn, 1975). Oligogalacturonates indeed exhibited a slight elicitor activity on *L. digitata* sporophytes (data not shown).

AOS Production Involves an NADPH Oxidase-Like Enzyme in *L. digitata* Sporophytes

Histochemical evidence based on the superoxide-specific stain NBT and the SOD inhibitor DDC (Fig. 5) points out that the AOS first formed during the

Table I. *Effects of various ionophores on the oxidative burst in L. digitata*

Compound	Function	Concentration	% Increase/Inhibition
		μM	%
Nonactin	K^+ ionophore	1	+33.7[a]
			−54.7[b]
Monensin	Na^+ ionophore	25	+61.4
		250	−99.3
2,4-DNP	H^+ ionophore	1	−75.5
		100	+1,184.6

[a] Without pre-incubation. [b] After 1-h pre-incubation.

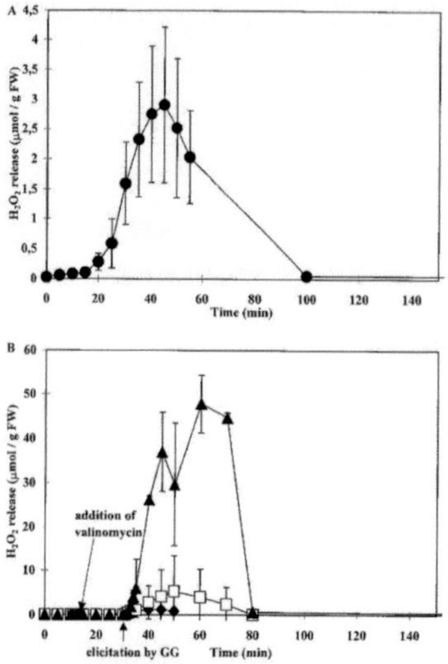

Figure 7. Effect of valinomycin on the oxidative burst triggered by oligoguluronates in *L. digitata*. A, Induction of an oxidative burst (without elicitation by GG) by 57 μM valinomycin (added at $t = 0$). B, Enhancement by valinomycin of the oxidative burst triggered by 150 μg mL^{-1} GG: dose-response behavior (control, ◆; 0.57 μM valinomycin, □; 5.7 μM valinomycin, ▲). Note that the addition of valinomycin enhances the oxidative burst, whereas concentrations around 57 μM are sufficient to trigger an oxidative burst without the physiological signal (GG).

oxidative burst is superoxide, most of which is subsequently converted to H_2O_2. This is further supported by the inhibition with DPI of the release of hydrogen peroxide (Fig. 3) and the staining with NBT (Fig. 5C) as diphenylene iodonium is known as an irreversible inhibitor of flavin-containing enzymes, including the superoxide-generating NAD(P) H oxidases of mammalian neutrophils (O'Donnell et al., 1993) and of higher plants (Auh and Murphy, 1995; Dwyer et al., 1996; Pugin et al., 1997). The fact that exogenous SOD did not affect the hydrogen peroxide release or NBT staining is likely due to a poor penetration of this enzyme through the thick brown algal cell wall.

The results obtained with quinacrine tend to further support that the oxidative-burst machinery of *L. digitata* contains a flavoprotein. Yet quinacrine is also a non-specific inhibitor of PLA_2 (Henderson et al., 1989) and consistent with the effect of chlorpromazine-HCl, this inhibitor may also indirectly affect AOS formation by interrupting a PLA_2-dependent transduction event. Altogether and keeping in mind the argument about the uses and misuses of inhibitors in studies of AOS sources in higher plants (Bestwick et al., 1999), we propose that a possible source of superoxide in *L. digitata* is an oxidase with a flavoprotein subunit (inhibited by DPI and likely by quinacrine).

Transduction of Oligoguluronate Signals Features Some Conserved Steps with Higher Plants

The general inhibitor of protein kinases, staurosporine, reported to block the protein phosphorylation events involved in induction of defense responses in tobacco cells upon elicitation by cryptogein (Viard et al., 1994) or oligogalacturonides (Mathieu et al., 1996), also inhibited the emission of H_2O_2 by *L. digitata* plantlets. Yet the protein phosphatase inhibitors calyculin A and cantharidin (Levine et al., 1994; Jabs et al., 1997) did not induce a release of H_2O_2 in *L. digitata*. Phospholipase A_2, another typical element of eukaryotic signal transduction pathways (Chapman, 1998), is also likely to participate in the oligoguluronate transduction chain, as suggested by the strong inhibition of H_2O_2 production by chlorpromazine-HCl, a selective inhibitor of this phospholipase following elicitation in higher plants (Chandra et al., 1996). In contrast no evidence was observed as to the involvement of G proteins in the transduction of oligoguluronate signals. Mastoparan, a specific activator of G proteins intrinsically capable of triggering an oxidative burst in soybean cells (Legendre et al., 1992), was ineffective to induce or enhance the production of AOS in *L. digitata* sporophytes, even at high concentrations.

Calcium entry, which is known to be involved with the activation of the oxidative machinery in higher plants (e.g. Jabs et al., 1997; Pugin et al., 1997) also appears to regulate the emission of AOS in *L. digitata*. This is demonstrated by the enhancing effect of the Ca^{2+} ionophore A23187 and the inhibitory effects of the antagonists of Ca^{2+} channels such as verapamil and methoxyverapamil (known to affect calcium channels in fucoid brown algae; Robinson, 1996), which completely blocked the oxidative burst.

Incubation of *L. digitata* sporophytes in the presence of oligoguluronates results in a marked potassium efflux (Fig. 8). Assuming that intracellular potassium concentrations are approximately 380 mM (see "Materials and Methods") and given the potassium concentration of the artificial seawater used in this study (500 μM instead of 10 mM in natural seawater), elicited *L. digitata* plantlets release about 2.6% of their internal potassium within the first h and about 13.1% overnight (14 h). Changes in plasma membrane permeability indeed appear to be directly involved in the regulation of oxidative burst in *L.*

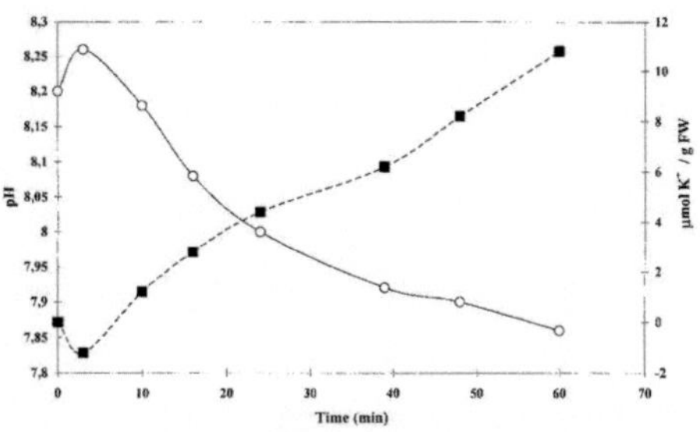

Figure 8. Time course of K^+ efflux (■) and pH variations (○) in the incubation medium of *L. digitata* plantlets (50 mg fresh weight mL^{-1}) treated at time 0 with 150 µg mL^{-1} oligoguluronates. The data reported here correspond to one representative experiment out of three.

digitata. The enormous enhancement of the effect of oligoguluronates by the K^+ ionophore valinomycin also points at the key role in the signal transduction chain of changes in membrane polarization by K^+ efflux. Yet the situation in *L. digitata* seems to be different from that in higher plants, where valinomycin fails to induce a burst when applied alone (Jabs et al., 1997). Pre-incubation with the uncoupler/H^+ carrier 2,4-dinitrophenol similarly enhanced the response, presumably also by changes in membrane polarization. Chloride channels may be involved too, as indicated by the partial inhibition of the burst by the anion channel antagonists, A9C, DIDS, and NPPB.

Establishing the hierarchical organization of the various transduction steps discussed above would now require an investigation of their links with other early events, such as ion fluxes, and the identification of defense responses that may function downstream or independently of the oxidative burst triggered by oligoguluronate recognition. The desensitizing effect of elicitation with alginate oligosaccharides suggests that such signaling pathway(s) cannot be repetitively activated over a certain period. A comparable refractory behavior during which cells remain insensitive to further elicitation for several hours is known in higher plants for various oligosaccharide elicitors, oligogalacturonides in tobacco (Rouet-Mayer et al., 1997) and soybean cells (Legendre et al., 1993), and chitin fragments in tomato cells (Felix et al., 1998).

Biological Significance of the Oxidative Burst in *L. digitata*

Enzymes that depolymerize alginate have been detected in a variety of marine molluscs, as well as in a number of marine bacteria associated with Laminariales, including *Pseudomonas alginovora*, a species epiphytic to *L. digitata* (Boyen et al., 1990), as well as strain H4 (Sawabe and Ezura, 1996) and *P. elyakovii* (Sawabe et al., 2000), two bacteria isolated from diseased cultures in *L. japonica* nurseries. Laminariales also are plagued by filamentous brown algal endophytes (Ellertsdottir and Peters, 1997), which are likely to express alginases when invading the host tissue. Therefore, recognition of alginate fragments by *L. digitata* sporophytes is likely to signal challenging with alginolytic, potentially detrimental organisms.

As reported for fungal pathogens of higher plants (Peng and Kùc, 1992), the oxidative burst may be viewed as an immediate, efficient defense in itself. The amounts of H_2O_2 released by *L. digitata* sporophytes upon recognition of oligoguluronates, up to 1 µmol g fresh weight^{-1}, together with the estimation that the apoplastic volume corresponds to approximately 10% of the tissue fresh weight in this brown alga (see Fig. 4), suggest that at the peak of production, the peroxide levels in the apoplasm can transiently reach concentrations as high as several millimolars, a figure comparable with those observed in soybean cells (Legendre et al., 1993). Such levels are at least one order of magnitude above the concentrations that were required to control the growth or even to kill the two kelp disease-associated bacteria investigated here, *P. elyakovii* and strain Ldm2. Exogenous applications of H_2O_2 in the mM range were indeed shown to control *Pseudoalteromonas teriolytica*, the causative agent of the red spot disease in *L. japonica* (Ezura et al., 1990; Sawabe et al., 1998).

Confocal microscopy (Fig. 4), histochemical staining (Fig. 5), as well as the comparison of algal fragments with and without the cortical cell layer (Fig. 6) indicate that reactivity to oligoguluronates is confined to cortical cells, i.e. cells that are the first to be exposed to pathogen attack. Young, fast-growing tissues of the blade are the most reactive upon elicitation. Yet meristematic areas showed almost no reaction. This suggests that the capability to produce AOS is linked to cell differentiation, as in cotton fibers (Potikha et al., 1999). The gametophyte generation, which in nature rapidly differentiates gametes and thus has a short life span, consistently does not recognize oligoguluronates as defense signals.

In conclusion, guluronate-rich alginate oligosaccharides, which are likely to be released during interactions with alginolytic organisms, elicit an oxidative burst in *L. digitata.* Since they are readily recognized by whole plantlets, oligoguluronates may thus be used as a new tool to decipher transduction pathways and defense reactions in Laminariales. This order belongs to the phylum of Heterokonta, which emerged as an independent lineage during the so-called crown diversification of higher eukaryotes (Bhattacharya et al., 1991). As in the case of the rhodophytes *Gracilaria conferta* (Weinberger et al., 1999) and *Chondrus crispus* (Bouarab et al., 1999), it would be of particular interest to delineate in *L. digitata* conserved ancestral pathways and machineries from those that are unique to the marine environment. In this respect the results reported here suggest that brown algal kelps feature some of the transduction cascade components known for higher plants, as well as a superoxide-generating oxidase with similarities to the well-characterized mammalian NADPH oxidase. The hydrogen peroxide concentrations generated in the surroundings of the algae appear sufficient to exert an inhibitory effect toward potentially harmful micro-organisms. Other defenses may involve the oxidation of intracellular iodide, leading to the release of toxic, iodinated compounds (Potin et al., 1999).

MATERIALS AND METHODS

Plant Material

Young *Laminaria digitata* plantlets (2–15 cm in length) were seeded and grown in a well-aerated, running-seawater tank as follows. Adult fertile sporophytes were placed for 1 week in the tank and allowed to release spores. Gametophytes developed along the walls and fertilization yielded young *L. digitata* plantlets. Cultures were grown using a photoperiod of 16 h of light and 8 h of darkness at a photon flux density of 50 to 60 μmol photons $m^{-2} s^{-1}$ and at temperatures corresponding to the surface seawater temperature at Roscoff, ranging from 9°C to 17°C throughout the seasonal cycle. Adult *L. digitata* sporophytes were collected by diving between Ile Verte and Ile de Batz, close to the Institute in Roscoff. In some experiments unialgal sporophytes were used, originating from the female gametophyte strain Lam dig Hel 1004 and the male gametophyte strain Lam dig Hel 1003 (kindly provided by D.G. Müller, Konstanz, Germany), and grown in the laboratory in Provasoli ES medium (Starr and Zeikus, 1987) prepared from artificial seawater (WIMEX, Krefeld, Germany, adjusted to 30‰ salinity). They were illuminated with daylight-type fluorescent lamps at an irradiance of 9 μE $m^{-2} s^{-1}$ for 10 h per day and kept at 10°C ± 1°C. Cultures were transferred to fresh medium at 1- or 2-week intervals.

Inhibition of the Growth of Kelp-Associated Bacterial Strains

One of the alginolytic strains used, isolated from diseased *L. japonica* and identified as *Pseudoalteromonas elyakovii,* was provided by T. Sawabe (Hokkaido University, Japan). The other one lacks taxonomic classification so far and was termed strain "Ldm2." It is an isolate from epiphytic bacteria living on decaying diseased young *L. digitata* collected from the field. Bacteria were first grown for around 12 h in ZoBell medium (ZoBell and Upham, 1944). During their exponential growth phase, 200 μL of the cultures were transferred as inocula to a mixture of 15 mL ZoBell medium and 10 mL sterile, filtered seawater with various additions of H_2O_2. Growth of the cultures was followed for up to 16 h by measuring the A_{600}.

Elicitors

Alginate oligosaccharides with a polymerization degree ranging from 15 to 25 (Heyraud et al., 1996) were prepared in the laboratory according to Haug et al. (1974) using sodium alginate from *Laminaria hyperborea* stipes (provided by B. Larsen, Trondheim University, Norway) and yielding three categories of alginate oligosaccharides (GG, MM, and MG blocks). In an alternate manner, alginate oligosaccharides were obtained by enzyme degradation of sodium alginate with an alginate lyase from the abalone *Haliotis tuberculata* (Boyen et al., 1990). In elicitation experiments, GG, MM, and MG blocks were applied at final concentrations ranging from 1 to 150 μg mL^{-1}.

Inhibitors and Activators of AOS Generation and Signal Transduction Chains

Several compounds were screened for their potential activating or inhibiting activity on signal transduction and AOS generation after elicitation: KCN (final concentration in the seawater eliciting medium of 50 μM; target, heme-dependent enzymes) and NaN_3 (5 and 50 μM; target, a wide range of redox enzymes) from stock solutions dissolved in H_2O, respectively; quinacrine (500 μM; target, flavin-dependent redox enzymes, in particular oxidases), DDC (1 mM; target, SOD), nifedipin (100 μM; target, Ca^{2+} channels), and chlorpromazine-HCl (10 and 260 μM, 1 mM; target, phospholipase A) from stock solutions dissolved in ethanol, respectively; and DPI (10 and 100 μM, 1 mM; target, NADPH oxidases) and A9C (100 μM and 1 mM; target, anion channels) from a stock in dimethyl sulfoxide (DMSO) and purchased from Sigma (L'Isle d'Abeau Chesnes, France). Calyculin A (10 μM; target, protein phosphatases) and cantharidin (10 μM; target, protein phosphatases) were from stocks in DMSO; staurosporine (5 μM; target, protein kinases), valinomycin (57 and 570 nM, 5.7 μM; target, K^+ ionophore), nonactin (1 μM; target, K^+ ionophore), monensin (25 and 250 μM; target, Na^+ ionophore), A23187 (5 and 50 μM; target, Ca^{2+} ionophore), DIDS (1 mM; target, anion channels), NPPB (100 μM; target, anion channels), verapamil (D200; 100 μM; target, Ca^{2+} channels), and methoxyverapamil (D600; 100 μM; target,

Ca^{2+} channels) in ethanol, respectively, were purchased from Calbiochem (France Biochem, Meudon, France).

Hydrogen Peroxide/AOS Measurements

The concentration of hydrogen peroxide present in the medium around plants was determined using the luminol chemiluminescence method (Glazener et al., 1991) with a luminometer (LUMAT LB 9507, EG&G Berthold, Bad Wildbach, Germany). It was usual that 400-μL aliquots (out of initially 10 mL) were taken for one measurement. In the luminometer, 50 μL of 20 units mL^{-1} of horseradish peroxidase (Boehringer Mannheim, Meylan, France, dissolved in pH 7.8 phosphate buffer) and 100 μL of 0.3 M luminol (5-amino-2, 3-dihydro-1,4-phthalazinedione, Sigma) solution were added to the sample. Chemiluminescence was recorded immediately after the last injection with a signal integration time of 10 s. For calculating the concentration of H_2O_2 present in the samples, calibration with a standard curve was carried out at least once during any series of measurements. Control assays of this method revealed that *L. digitata* sporophytes and gametophytes maintain very low levels (<100 nM) of H_2O_2 in their surroundings under steady-state conditions. Catalase at 100 U mL^{-1} and SOD at 200 U mL^{-1} (Boehringer Mannheim) were used to test the types of AOS detected in the assay.

O_2 Measurements

Oxygen consumption was measured in darkness with an oxygen electrode (DW3 Clark-type, Hansatech Instruments, King's Lynn, UK). The 20-mL reaction vessel had a small opening allowing for the injection of microliter volumes of solutions (e.g. containing oligosaccharides) during the course of the experiments. Young *L. digitata* sporophytes approximately 3 cm in length and 50 mg in fresh weight were incubated in natural seawater to determine respiration rates.

Histochemical Staining for O_2^- Detection

Immediately after triggering an oxidative burst by elicitation (including in the presence of the SOD inhibitor DDC, 200 U mL^{-1} of SOD, or 10 μM DPI, respectively), plantlets were placed in 10 mM NaN_3 in 10 mM potassium phosphate buffer (pH 7.8) and immersed for 15 to 30 min in 3 mL of the same buffer containing 0.1% (w/v) NBT (Sigma) as described in Jabs et al. (1996). Plant pigments were subsequently cleared by boiling for 2 min in alcoholic lactophenol (95% [w/v] ethanol:lactophenol, 2:1). Macroscopic and microscopic photographs were taken of entire plantlets and of thin cross sections to localize superoxide formation.

Laser Scanning Confocal Microscopy

DCFH-DA was dissolved in DMSO or ethanol to produce 10 mM stock, which was frozen as aliquots. Young *L. digitata* thalli were cross-sectioned using razor blades, and cross sections were placed in the dark for 15 min into Petri dishes containing 5 mL of 22 μm-filtered seawater and 5 μL of DCFH-DA from the stock solution. Petri dishes 35 mm in diameter were prepared for microscopy by sealing a glass coverslip over a 18-mm hole drilled in the bottom of the dish. Cross sections were removed from the DCFH-DA-loading solution, rinsed with 22 μm filtered seawater, and affixed to the bottom of a Petri dish containing 2 mL of seawater for microscopy. Imaging was performed using a laser scanning confocal microscope (model IX 70/Fluoview, Olympus, Tokyo) with images recorded every 60 s. An argon/krypton OMNICHROM laser was used for excitation at 488 nm set on 20% power, with 525 nm emission (channel 1) and 580 nm emission (channel 2). Images were captured over a 15-min time course. Thalli were elicited 5 min after the beginning of the time course with 150 μg mL^{-1} oligoguluronates. Analysis of images was performed using the Fluoview software (version 2.0, Olympus).

K^+ and H^+ Fluxes

The protocol of Mathieu et al. (1991) for measurement of potassium efflux was modified as follows. The extracellular K^+ concentration was determined at intervals using a K^+-specific electrode (F2312K, Radiometer, Copenhagen). Artificial seawater (450 mM NaCl, 10 mM $CaCl_2$, 5 mM $MgCl_2$, 2 mM $NaHCO_3$, and 40 mM $MgSO_4$) adjusted to 500 μM K^+ was used for all assays with living plants. A pH electrode was constantly inserted in the medium surrounding the plantlets. For comparability, readings were taken at every measurement of K^+. The intracellular K^+ concentration of *L. digitata* plantlets was determined by drying 10 samples of plantlets, which were subsequently analyzed by means of induction-coupled plasma-atom emission spectroscopy by the Service Central d'Analyse of Centre National de la Recherche Scientifique (Vernaison, France). The average potassium content of dried plantlets was 8.771% $\pm$ 1.814%. Given that, in *L. digitata*, 1 g of dry weight corresponds to approximately 6.56 g fresh weight (this study) and that the apoplastic volume represents around 10% of the tissue fresh weight (Mabeau and Kloareg, 1987), this corresponds to intracellular K^+ concentrations around 380 mM. This compares well with the 300 mM measured for other brown algae (Reed and Barron, 1983).

ACKNOWLEDGMENTS

We would like to thank Claire Gachon (Ecole Normale Supérieure-Ulm, Paris), Jean-François Lennon, Aldo Asensi, Guy Levavasseur, Sylvie Rousvoal and Christophe Richard (CNRS, Roscoff, France), Alain Heyraud (CNRS, Grenoble, France), Dieter G. Müller (University of Konstanz), and J.-M. Legendre (Centre Hospitalier Universitaire Morvan, Brest, France) for their help and G. Stark (University of Konstanz) for useful suggestions.

Received April 28, 2000; accepted August 15, 2000.

LITERATURE CITED

Allan AC, Fluhr R (1997) Two distinct sources of elicited reactive oxygen species in tobacco epidermal cells. Plant Cell **9:** 1559–1572

Apostol I, Low PS, Heinstein P, Stipanovic RD, Altman DW (1987) Inhibition of elicitor-induced phytoalexin formation in cotton and soybean cells by citrate. Plant Physiol **84:** 1276–1280

Auh C-K, Murphy TM (1995) Plasma membrane redox enzyme is involved in the synthesis of O_2^- and H_2O_2 by *Phytophthora* elicitor-stimulated rose cells. Plant Physiol **107:** 1241–1247

Baker CJ, Orlandi EW (1995) Active oxygen in plant pathogenesis. Annu Rev Phytopathol **33:** 299–321

Baldridge CW, Gerard RW (1933) The extra respiration of phagocytosis. Am J Physiol **103:** 235–236

Bestwick C, Bolwell P, Mansfield J, Nicole M, Wojtaszek P (1999) Generation of the oxidative burst: scavenging for the truth. Trends Plant Sci **4:** 88–89

Bhattacharya D, Stickel SK, Sogin ML (1991) Molecular phylogenetic analysis of actin genic regions from *Achlya bisexualis* (Oomycota) and *Costaria costata* (Chromophyta). J Mol Evol **33:** 525–536

Boller T (1995) Chemoperception of microbial signals in plant cells. Annu Rev Plant Physiol Plant Mol Biol **46:** 189–214

Bolwell GP, Butt VS, Davies DR, Zimmerlin A (1995) The origin of the oxidative burst in plants. Free Radic Res **23:** 517–532

Bolwell GP, Davies RD, Gerrish C, Auh C-K, Murphy TM (1998) Comparative biochemistry of the oxidative burst produced by rose and French bean cells reveals two distinct mechanisms. Plant Physiol **116:** 1379–1385

Bouarab K, Potin P, Correa J, Kloareg B (1999) Sulfated oligosaccharides mediate the interaction between a marine red alga and its green algal pathogenic endophyte. Plant Cell **11:** 1635–1650

Boudart G, Lafitte C, Barthe JP, Frasez D, Esquerré-Tugayé M-T (1998) Differential elicitation of defense responses by pectic fragments in bean seedlings. Planta **206:** 86–94

Boyen C, Kloareg B, Polne-Fuller M, Gibor A (1990) Preparation of alginate lyases from marine molluscs for protoplast isolation in brown algae. Phycologia **29:** 173–181

Brisson LF, Tenhaken R, Lamb C (1994) Function of oxidative cross-linking of cell wall structural proteins in plant disease resistance. Plant Cell **6:** 1703–1712

Cazalé A-C, Rouet-Mayer M-A, Barbier-Brygoo H, Mathieu Y, Laurière C (1998) Oxidative burst and hypoosmotic stress in tobacco cell suspensions. Plant Physiol **116:** 659–669

Chandra S, Heinstein P, Low PS (1996) Activation of phospholipase A by plant defense elicitors. Plant Physiol **110:** 979–986

Chapman KD (1998) Phospholipase activity during plant growth and development and in response to environmental stress. Trends Plant Sci **3:** 419–426

Chen Z, Silva H, Klessig DF (1993) Active oxygen species in the induction of plant systemic acquired resistance by salicylic acid. Science **262:** 1883–1886

Collén J, Pedersén M (1994) A stress-induced oxidative burst in *Eucheuma platycladum* (Rhodophyta). Physiol Plant **92:** 417–422

Dempsey DA, Klessig DF (1994) Salicylic acid, active oxygen species and systemic acquired resistance in plants. Trends Cell Biol **4:** 334–338

Desikan R, Hancock JT, Coffey MJ, Neill SJ (1996) Generation of active oxygen in elicited cells of *Arabidopsis thaliana* is mediated by a NADPH oxidase-like enzyme. FEBS Lett **382:** 213–217

Doke N (1983a) Involvement of superoxide anion generation in the hypersensitive response of potato tuber tissues to infection with an incompatible race of *Phytophthora infestans* and to the hyphal wall components. Physiol Plant Pathol **23:** 345–357

Doke N (1983b) Generation of superoxide anion by potato tuber protoplasts during the hypersensitive response to hyphal wall components of *Phytophthora infestans* and specific inhibition of the reaction by suppressors of hypersensitivity. Physiol Plant Pathol **23:** 359–367

Dumas B, Sailland A, Cheviet JP, Freyssinet G, Pallett K (1993) Identification of barley oxalate oxidase as a germin-like protein. C R Acad Sci **316:** 793–798

Dwyer SC, Legendre L, Low PS, Leto TL (1996) Plant and human neutrophil oxidative burst complexes contain immunologically related proteins. Biochim Biophys Acta **1289:** 231–237

Ebel J (1998) Oligoglucoside elicitor-mediated activation of plant defense. Bioessays **20:** 569–576

Ebel J, Bhagwat AA, Cosio EG, Freger M, Kissel U, Mithöfer A, Waldmüller T (1995) Elicitor-binding proteins and signal transduction in the activation of a phytoalexin defense response. Can J Bot Suppl **73:** S506–S510

Ellertsdottir E, Peters AF (1997) High prevalence of infection by endophytic brown algae in populations of *Laminaria* spp. (Phaeophyceae). Marine Ecol Prog Series **146:** 135–143

Ezura Y, Hara Y, Kimura T (1990) A control method for the red-spot injury occurring on seed twines for the cultivation of Makonbu *Laminaria japonica*. Nippon Suisan Gakkashi **56:** 2045–2051

Felix G, Baureithel K, Boller T (1998) Desensitization of the perception system for chitin fragments in tomato cells. Plant Physiol **117:** 643–650

Frahry G, Schopfer P (1998) Inhibition of O_2^--reducing activity of horseradish peroxidase by diphenyleneiodonium. Phytochemistry **48:** 223–227

Fritig B, Heitz T, Legrand M (1998) Antimicrobial proteins in induced plant defense. Curr Opin Immunol **10:** 16–22

Fry SC, Aldington S, Hetherington PR, Aitken J (1993) Oligosaccharides as signals and substrates in the plant cell wall. Plant Physiol **103:** 1–5

Glazener JA, Orlandi EW, Harmon GL, Baker CJ (1991) An improved method for monitoring active oxygen in bacteria-treated suspension cells using luminol-dependent chemiluminescence. Physiol Mol Plant Pathol **39:** 123–133

Groom QJ, Torres MA, Fordham-Skelton AP, Hammond-Kosack KE, Robinson NJ, Jones JDG (1996) *rbohA*, a rice homolog of the mammalian *gp91phox* respiratory burst oxidase gene. Plant J **10:** 515–522

Hammond-Kossack KE, Jones JDG (1996) Resistance gene-dependent plant defense responses. Plant Cell **8:** 1773–1791

Haug A, Larsen B, Smidsrød O (1974) Uronic acid sequence in alginate from different sources. Carbohydrate Res **32:** 217–225

Henderson LM, Chappell JB, Jones OT (1989) Superoxide generation is inhibited by phospholipase A2 inhibitors: role for phospholipase A2 in the activation of the NADPH oxidase. Biochem J **264:** 249–255

Heyraud A, Gey C, Leonard C, Rochas C, Girond S, Kloareg B (1996) NMR spectroscopy analysis of oligoguluronates and oligomannuronates prepared by acid or enzymatic hydrolysis of homopolymeric blocks of alginic acid: application to the determination of the substrate specificity of *Haliotis tuberculata* alginate lyase. Carbohydrate Res **289:** 11–23

Jabs T, Dietrich RA, Dangl JL (1996) Initiation of runaway cell death in an *Arabidopsis* mutant by extracellular superoxide. Science **273:** 1853–1856

Jabs T, Tschöpe M, Colling C, Hahlbrock K, Scheel D (1997) Elicitor-stimulated ion fluxes and O_2^- from the oxidative burst are essential components in triggering defense gene activation and phytoalexin synthesis in parsley. Proc Natl Acad Sci USA **94:** 4800–4805

Keller T, Dalmude HG, Werner D, Doerner P, Dixon RA, Lamb C (1998) A plant homolog of the neutrophil NADPH oxidase $gp91^{phox}$ subunit gene encodes a plasma membrane protein with Ca^{2+} binding motifs. Plant Cell **10:** 255–266

Kloareg B, Quatrano R (1988) Structure of the cell walls of marine algae and ecophysiological function of the matrix polysaccharides. Annu Rev Ocean Marine Biol **26:** 259–315

Kohn R (1975) Ion binding on polyuronates. Pure Appl Chem **42:** 371–397

Küpper FC, Schweigert N, Ar Gall E, Legendre JM, Vilter H, Kloareg B (1998) Iodine uptake in Laminariales involves extracellular, haloperoxidase-mediated oxidation of iodide. Planta **207:** 163–171

Lamb C, Dixon RA (1997) The oxidative burst in plant disease resistance. Annu Rev Plant Physiol Plant Mol Biol **48:** 251–275

Legendre L, Heinstein PF, Low PS (1992) Evidence for participation of GTP-binding proteins in elicitation of the rapid oxidative burst in cultured soybean cells. J Biol Chem **267:** 20140–20147

Legendre L, Rueter S, Heinstein PF, Low PS (1993) Characterization of the oligogalacturonide-induced oxidative burst in cultured soybean (*Glycine max*) cells. Plant Physiol **102:** 233–240

Levine A, Tenhaken R, Dixon R, Lamb C (1994) H_2O_2 from the oxidative burst orchestrates the plant hypersensitive disease resistance response. Cell **79:** 583–593

Lurin C, Geelen D, Barbier-Brygoo H, Guern J, Maurel C (1996) Cloning and functional expression of a plant voltage-dependent chloride channel. Plant Cell **8:** 701–711

Mabeau S, Kloareg B (1987) Isolation and analysis of the cell walls of brown algae: *Fucus spiralis, F. ceranoides, F. vesiculosus, F. serratus, Bifurcaria bifurcata* and *Laminaria digitata*. J Exp Bot **38:** 1573–1580

Martinez C, Montillet JL, Bresson E, Agnel JP, Dai GH, Daniel JF, Geiger JP, Nicole M (1998) Apoplastic peroxidase generates superoxide anions in cells of cotton cotyledons undergoing the hypersensitive reaction to *Xanthomonas campestris* pv. *malvacearum* race 18. Mol Plant-Microbe Interact **11:** 1038–1047

Mathieu Y, Kurkdijan A, Xia H, Guern J, Koller A, Spiro MD, O'Neill M, Albersheim P, Darvill A (1991) Membrane responses induced by oligogalacturonides in suspension-cultured tobacco cells. Plant J **1:** 333–343

Mathieu Y, Lapous D, Thomine S, Laurière C, Guern J (1996) Cytoplasmic acidification as an early phosphorylation-dependent response of tobacco cells to elicitors. Planta **199:** 416–424

Morel F, Doussiere J, Vignais PV (1991) The superoxide-generating oxidase of phagocytic cells. Eur J Biochem **201:** 523–546

O'Donnell V, Tew DG, Jones OTG, England PJ (1993) Studies on the inhibitory mechanism of iodonium compounds with special reference to neutrophil NADPH oxidase. Biochem J **290:** 41–49

Otte O, Barz W (1996) The elicitor-induced oxidative burst in cultured chickpea cells drives the rapid insolubilization of two cell wall structural proteins. Planta **200:** 238–246

Peng M, Kùc J (1992) Peroxidase-generated hydrogen peroxide as a source of antifungal activity in vitro and on tobacco leaf disks. Phytopathology **82:** 696–699

Potikha TS, Collins CC, Johnson DI, Delmer DP, Levine A (1999) The involvement of hydrogen peroxide in the differentiation of secondary walls in cotton fibers. Plant Physiol **119:** 849–858

Potin P, Bouarab K, Küpper FC, Kloareg B (1999) Oligosaccharide signalling in marine plant-microbe interactions. Curr Opin Microbiol **2:** 276–83

Pugin A, Frachisse JM, Tavernier E, Bligny R, Gout E, Douce R, Guern J (1997). Early events induced by the elicitor cryptogein in tobacco cells: involvement of a plasma membrane NADPH oxidase and activation of glycolysis and the pentose phosphate pathway. Plant Cell **9:** 2077–2091

Reed RH, Barron JA (1983) Physiological adaptation of *Pilayella littoralis* from marine and estuarine sites. Bot Marina **26:** 409–416

Richberg MH, Aviv DH, Dangl JL (1998) Dead cells do tell tales. Curr Opin Plant Biol **1:** 480–485

Robinson KR (1996) Calcium and the photopolarization of *Pelvetia* zygotes. Planta **198:** 378–384

Rouet-Mayer M-A, Mathieu Y, Cazalé A-C, Guern J, Laurière C (1997) Extracellular alkalinization and oxidative burst induced by fungal pectin lyase in tobacco cells are not due to perception of oligogalacturonide fragments. Plant Physiol Biochem **35:** 321–330

Sawabe T, Ezura Y (1996) Regeneration of *Laminaria japonica* Areschoug (Laminariales, Phaeophyceae) protoplasts isolated with bacterial alginase. Plant Cell Rep **15:** 892–895

Sawabe T, Makino H, Tatsumi M, Nakano K, Tajima K, Iqbal MM, Yumoto I, Ezura Y, Christen R (1998) *Pseudoalteromonas bacteriolytica* sp. nov., a marine bacterium that is the causative agent of red spot disease in *Laminaria japonica*. Int J System Bacteriol **48:** 769–774

Sawabe T, Tanaka R, Iqbal MM, Tajima K, Ezura Y, Ivanova EP, Christen R (2000) Assignment of *Alteromonas elyakovii* KMM 162^T and five strains from spot-wounded fronds of *Laminaria japonica* to *Pseudoalteromonas elyakovii* comb. nov. and the extended description of the species. Int J Syst Ecol Microbiol **50:** 265–271

Scheel D (1998) Resistance response physiology and signal transduction. Curr Opin Plant Biol **1:** 305–310

Scheel D, Parker JE (1990) Elicitor recognition and signal transduction in plant defense gene activation. Zeitschrift für Naturforschung **45:** 569–575

Starr RC, Zeikus JA (1987) UTEX: the culture collection of algae at the University of Texas at Austin. J Phycol Suppl **23:** 1–47

Tavernier E, Wendehenne D, Blein JP, Pugin A (1995) Involvement of free calcium in action of cryptogein, a proteinaceous elicitor of hypersensitive reaction in tobacco cells. Plant Physiol **109:** 1025–1031

Tenhaken R, Levine A, Brisson LF, Dixon RA, Lamb C (1995) Function of the oxidative burst in hypersensitive disease resistance. Proc Natl Acad Sci USA **92:** 4158–4163

Thordal-Christensen H, Zhang Z, Wei Y, Collinge DB (1997) Subcellular localization of H_2O_2: H_2O_2 accumulation in papillae and hypersensitive response during the barley-powdery mildew interaction. Plant J **11:** 1187–1194

Torres MA, Onouchi H, Hamada H, Machida C, Hammond-Kosack KE, Jones JDG (1998). Six *Arabidopsis thaliana* homologues of the human respiratory burst oxidase (*gp91phox*). Plant J **14:** 365–370

Van Camp WV, Van Montagu M, Inzé D (1998). H_2O_2 and NO: redox signals in disease resistance. Trends Plant Sci **3:** 330–334

Vardi A, Berman-Frank I, Rozenberg T, Hadas O, Kaplan A, Levine A (1999) Programmed cell death of the dinoflagellate *Peridinium gatunense* is mediated by CO_2 limitation and oxidative stress. Curr Biol **9:** 1061–1064

Viard M-P, Martin F, Pugin A, Ricci P, Blein J-P (1994) Protein phosphorylation is induced in tobacco cells by the elicitor cryptogein. Plant Physiol **104:** 1245–1249

Weinberger F, Friedlander M, Hoppe H-G (1999) Oligoagars elicit an oxidative burst in *Gracilaria conferta* (Rhodophyta). J Phycol **35:** 747–755

Wojtaszek P (1997) Oxidative burst: an early plant response to pathogen infection. Biochem J **322:** 681–692

Zhang Z, Collinge DB, Thordal-Christensen H (1995) Germin-like oxalate oxidase, a H_2O_2-producing enzyme, accumulates in barley attacked by the powdery mildew fungus. Plant J **8:** 139–145

ZoBell CE, Upham HC (1944) A list of marine bacteria including descriptions of sixty new species. Bull Scripps Inst Ocean Univ Calif (Tech Ser) **5:** 239–292

7

Planta (1998) 207: 163–171

Planta

Iodine uptake in Laminariales involves extracellular, haloperoxidase-mediated oxidation of iodide

F.C. Küpper[1], N. Schweigert[1], E. Ar Gall[1], J.-M. Legendre[2], H. Vilter[3], B. Kloareg[1]

[1]Centre d'Etudes Océanologiques et de Biologie Marine, CNRS, UPR 9042, F-29682 Roscoff, Brittany, France
[2]Service de Médecine Nucléaire, CHU Morvan, F-29200 Brest, Brittany, France
[3]Zurmaiener Straße 16, D-54292 Trier, Germany

Received: 11 February 1998 / Accepted: 18 June 1998

Abstract. Sporophytes of *Laminaria digitata* (L.) Lamour. were assayed for their content of accumulated iodine, which ranged from 0.4% of dry weight in adult plants up to 4.7% for young plantlets. Sporophyte tissue from *Laminaria saccharina* (L.) Lamour. and *L. digitata* took up iodide according to Michaelis-Menten kinetics. Hydrogen peroxide and various substances known to interfere with oxidative metabolism were shown to either inhibit or enhance the uptake of iodide, confirming that apoplastic oxidations play a key role in iodide uptake in *Laminaria*. Consistently, iodide uptake was triggered in *L. saccharina* protoplasts by incubation in the presence of hydrogen peroxide. Similarly, the uptake of iodide was enhanced in *L. digitata* gametophytes by addition of haloperoxidase, suggesting that this enzyme catalyses the oxidation of iodide by hydrogen peroxide and plays a key role in iodine uptake. Oxidative stress resulted in a marked efflux of the intracellular iodine. In both influx and efflux experiments, a marked proportion (10–30%) of the tracer was not accounted for, indicating volatilisation of iodine. The mechanism and possible functions of the accumulation of iodine by kelps are discussed.

Key words: Apoplastic oxidative metabolism – Haloperoxidase – Hydrogen peroxide – Iodine – *Laminaria*

Introduction

The total iodine concentration in the open ocean ranges between 0.44 and 0.49 μM. The major iodine species in coastal sea waters are iodate (IO_3^-) and iodide (I^-), along with smaller fractions of molecular iodine (I_2), hypoiodous acid (HIO) as well as various iodinated organic compounds (Truesdale et al. 1995). In the euphotic layer, iodide concentrations can reach as much as 50% of the total inorganic iodine, ca. 0.25 μM (Wong 1977). Brown algae from the order Laminariales accumulate iodine to more than 30,000 times the concentration of this element in seawater, up to levels as high as 1% of dry weight in *Laminaria digitata*, depending on wave exposure, geographic location, and depth (Black 1948, Young and Langille 1958; Saenko et al. 1978). Iodine was actually discovered in kelps by Courtois (see Shaw 1962) and Laminariales still remain a major source for the recovery of this element, mainly from the mariculture of *Laminaria japonica* in China (Wu and Lin 1987).

Understanding the biochemical pathways of iodine uptake in Laminariales has evoked interest since as early as the beginning of the twentieth century. Dangeard (1928) and Kylin (1929) first established the link between oxidative metabolism and the uptake of iodine. They discovered that the natural release of free iodine (I_2) from kelps, a process termed "iodovolatilisation", requires oxygen and takes place in the apoplast of the thallus surface cells. Since then, iodovolatilisation has been considered to account for the higher iodine content of the air in coastal areas (Dangeard 1957). Kylin (1929) postulated the involvement of an "iodide oxidase" in the thallus surface mucilage and discussed the oxidation of iodide prior to uptake, suggesting that I_2 was the species finally taken up. He presumed that, inside the cells, I_2 (0) would again be reduced to I^- (−1). Given the parallelism between iodine uptake kinetics by *L. digitata* and the speciation between molecular iodine and its aqueous form, Shaw (1959) proposed that HIO (+1) is the actual iodine form taken up in Laminariales.

The availability of radioactive ^{131}I allowed for accurate quantitative measurements of iodine uptake in brown algae (Roche and Yagi 1952). Uptake mechanisms were studied in the Fucales *Ascophyllum nodosum* (Baily and Kelly 1955) and *Fucus ceranoides* (Klemperer 1957), showing Michaelis-Menten kinetics and indicating that iodide and not iodate was the iodine

Abbreviations: AOS = activated oxygen species; DNP = 2,4-dinitrophenol; GOD = glucose oxidase; VHCs = volatile halogenated compounds

Correspondence to: B. Kloareg; E-mail: kloareg@sb-roscoff.fr; Fax: 33 (2) 98292324

species involved in the uptake. Based on the inhibition of iodine transport by various reducing agents in the kelp *Nereocystis luetkeana*, Tong and Chaikoff (1955) confirmed the dependence of iodine accumulation on oxygen and suggested involvement of hydrogen peroxide H_2O_2 in the oxidation of iodide. The radioactive tracer was detected in the tissue as inorganic iodine and mono- and di-iodotyrosine. These findings were later confirmed in *L. flexicaulis* (now *L. digitata*) by André (1965, 1971). No kinetic data are available for the accumulation of iodine in kelps, however, in spite of their marked ability to take up and concentrate iodine.

The presence of an alginate-bound peroxidase in *L. digitata* was reported by Murphy and O hEocha (1973b), who suggested its involvement in the oxidation of iodide as well as in the formation of iodoamino acids. More recently, non-heme vanadium (V)-containing haloperoxidases were discovered in the cell walls of brown algae (Vilter et al. 1983; Vilter 1984), including in the Laminariales (e.g., de Boer et al. 1986; Butler et al. 1990; Jordan and Vilter 1991). Haloperoxidases catalyse the oxidation of bromide and iodide into hypohalides. It was therefore suggested that, as catalysts of the oxidation of iodide into hypoiodous acid, haloperoxidases are involved in the accumulation of iodide in brown algae (Murphy and O hEocha 1973a; Vilter et al. 1983). However, no experimental evidence supporting involvement of haloperoxidase in iodine uptake has been reported so far.

In this paper, we report the kinetics of iodine accumulation by sporophytic tissues of *L. saccharina* and *L. digitata* and show that iodide is taken up according to a facilitated diffusion mechanism based on the oxidation of iodide in the apoplast. In addition, we demonstrate that the uptake of iodide by *L. digitata* gametophytes is markedly enhanced by incubation in the presence of haloperoxidase.

Materials and methods

Plant material and protoplast isolation. Laminaria saccharina (L.) Lamour and *Laminaria digitata* (L.) Lamour sporophytes were collected in the vicinity of Roscoff (Brittany, France) between October 1994 and July 1996. Sporophytes were kept in tanks of running sea water and used within 2 d after collection. For the experiments involving direct assessment of sporophyte tissue, discs, 1.7 cm in diameter, were punched out of the lower blade margins and allowed to recover in natural seawater for about 1 h. The discs averaged 58 mg and 220 mg in fresh weight for *L. saccharina* and *L. digitata*, respectively. Gametophyte cultures of *L. digitata* were established and maintained as described previously (Ar Gall et al. 1996).

Protoplasts were isolated from the meristematic regions and lower margins of *L. saccharina* blades, according to Benet et al. (1997). Yields ranged between 3×10^7 and 5×10^7 protoplasts · (g FW)$^{-1}$. Protoplasts were resuspended in artificial sea water (ASW) brought to an osmotic pressure of 1250 mosmol with KCl and $MgCl_2$ (ASW_{1250}: 400 mM NaCl, 80 mM $MgCl_2$, 22 mM $MgSO_4$, 85 mM KCl, 7 mM $CaCl_2$, 10 mM Hepes, pH adjusted to 8.2 with NaOH). Cell density was adjusted to 5×10^6 cells · ml^{-1} and the suspension was stored overnight in the dark at 12 °C on a slowly revolving rotary shaker. Iodine-uptake experiments were carried out within 24 h of protoplast isolation. Depending on the conditions of the experiment, cell densities were adjusted with ASW_{1250} to 2.5×10^6–2.5×10^5 cells · ml^{-1}. A control experiment showed that the use of ASW_{1250} instead of natural seawater resulted in no differences in iodine uptake.

Determination of accumulated iodine in L. digitata. L. digitata sporophytes were sampled in the sublittoral off Enez Glas near Roscoff and assayed for accumulated iodine as follows. Tissue disks were punched out of the blade with a cork borer, dried extensively at 60 °C and powdered with a Waring blender. Total iodine content was then determined by neutron activation analysis at the Département d'Analyse Élémentaire, Service Central d'Analyse, Centre National de la Recherche Scientifique (Vernaison, France).

Measurements of iodine uptake in Laminaria tissue and protoplasts. Tissue discs, gametophytes, or protoplasts were incubated at 10–15 °C on a rotary shaker in the presence of autoclaved, natural or artificial seawater, supplemented with various concentrations of KI. The amount of iodide contributed by the various salts in ASW was estimated at ca. 2 μM from the information from the suppliers regarding the contaminant trace elements. $Na^{123}I$ (supplied by CIS Bio International, Gif-sur-Yvette, France) or $Na^{125}I$ (supplied by Amersham, UK) was added to the incubation medium, resulting in a specific activity of 15 kBq · ml^{-1}(^{123}I) or 940 Bq · ml^{-1}(^{125}I), respectively, and with no significant change in iodide concentration. Biomass was designed so that depletion of radioactive iodine from the incubation medium at the end of the experiments was less than 20% of the initial radioactivity. At the end of incubation, 1-ml aliquots of the supernatant liquids were taken and tissue discs were rinsed exhaustively in seawater and blotted dry on a paper towel. The radioactivity present in the liquid and discs was determined with a gamma counter, consisting of a 7.6 × 7.6 cm well type NaI (Tl) crystal (Bicron), connected to a multichannel analyser (Canberra, France). Unless mentioned otherwise, measurements were performed in five independent replicates. The proportion of volatilised iodine was estimated by subtracting from the initial supply both the amount taken up by the tissue and that remaining in the incubation medium.

Protoplast or gametophyte suspensions were centrifuged for 3 min at 1,500 rpm using a Jouan GR 4.11 centrifuge, washed in ASW_{1250}, pelleted again and the radioactivity remaining in the pellet was determined by gamma counting. When using ^{123}I, data were corrected for the loss of tracer activity during the time course of experiments. Carry-over into protoplasts from the incubation medium was estimated in a series of control experiments, using [^{14}C] sorbitol (Amersham). The carry-over of radioactivity after 1 h incubation and one wash cycle was less than 0.6% of the tracer present in the original incubation medium and did not change significantly upon incubation in the presence of H_2O_2 or glucose oxidase (GOD)-glucose, whilst iodine accumulation in comparable protoplast suspensions amounted to about 10% of the initial supply within 10 min.

Monitoring of iodine efflux by L. digitata plantlets. Laminaria digitata plantlets, ca. 5 cm in size and with an average iodine content of 4.7% of dry weight (as described below), were charged for 24 h with ca. 37 kBq of iodine each (in volumes of 200 ml seawater). The radioactive incubation medium was then removed and plantlets were rinsed and left in 200 ml seawater for another 12–15 h, allowing for the determination under normal conditions of the release of accumulated iodine. Plantlets were finally transferred into 5 ml seawater and iodine efflux upon addition of 50 μM, 200 μM or 2 mM exogenous H_2O_2 was monitored by taking 400-μl aliquots at intervals of 5, 10, 15, 30, 60, 120 and 180 min after the onset of the stress. Measurements of the radioactivity remaining in the plantlets were also carried out, in order to compare the total amount of iodine released from the sporophytes with that recovered in the incubation medium.

Regulation of iodine uptake by peroxide and haloperoxidase. For studying the effect of H_2O_2 on iodine uptake, *Laminaria* samples were either directly provided with dilutions of a H_2O_2 stock solution

(Merck) or incubated in the presence of GOD (Sigma) and 100 mM glucose. Addition of 100 mM glucose alone did not result in any stimulatory effect on the uptake of iodine. The actual amount of H_2O_2 generated with the GOD-glucose system was assayed in seawater using the luminol chemiluminescence method (Glazener et al. 1991) and a Berthold LB 9507 luminometer. Hydrogen peroxide concentrations after 10 min were 63, 200, 910 µM for GOD activities of 0.025, 0.1 and 1.0 to 2.5 U · ml^{-1}, respectively. Steady-state, permanent concentrations of 5 µM H_2O_2 were generated in *L. digitata* gametophyte suspensions by using 0.025 U · ml^{-1} GOD, 100 mM glucose and 6.4 U · ml^{-1} catalase.

A variety of chemical compounds known to interfere with oxidative metabolism was tested for possible inhibitory or enhancing effects on iodine uptake in tissue (see Table 1 for details). Unless mentioned otherwise, they were added to the tissue pieces in natural seawater 1 h prior to incubation with the radioactive tracer. The putative inhibitory effect on iodine uptake of 2,4-dinitrophenol (DNP; Amat, 1985) was investigated by preincubating protoplasts and tissue discs for 15, 30 and 60 min, with uncoupler concentrations ranging from 10 µM to 1 mM.

To address the possible involvement of haloperoxidase in iodine uptake, sporophyte tissue discs proved an unsuitable experimental system because of their high, yet heterogenous intrinsic rates of iodine uptake. *Laminaria digitata* gametophytes were therefore assayed for uptake of iodide in the presence of either or both H_2O_2 and haloperoxidase. Two different vanadium-(V)-dependent haloperoxidases were tested: a partially purified enzyme from *Ascophyllum nodosum*, referred to as A.n. I, a cell wall isoform from the transitional region between cortex and medulla (Vilter 1994); and the cortical isoenzyme from *L. digitata* (Jordan et al. 1991), prepared from freshly collected fronds according to Vilter (1994). Gametophytes (35 mg FW on average in 2 ml culture medium) were incubated in natural seawater supplemented with 10 µM KI. Controls consisted of 20-min incubations, without any additions or in the presence of 10 µM peroxide only, or with the steady-state concentration of 5 µM. The effect of haloperoxidase on iodine uptake was assessed in similar oxidative conditions, yet with addition of 10 U · m^{-1} of A.n. I haloperoxidase or of 6 U · ml^{-1} of haloperoxidase from *L. digitata*. At the end of incubation, gametophytes were washed and centrifuged three times, the amount of accumulated iodine was determined and the uptake of label was compared with the amount removed from the liquid. Experiments were performed in 5–10 replicates.

Table 1. Pharmacological investigation of the relationships between iodine uptake and oxidative metabolism. Tissue disks were preincubated for 1 h in the presence of various compounds known to interfere with the apoplastic concentration of activated oxygen species. Subsequent inhibition (–) or enhancement (+) of the rates of iodine uptake (10 µM KI, 10 min) are reported as percent of the control sporophyte tissue. Ld, *Laminaria digitata*; Ls, *L. saccharina*

Compound or Enzyme	Concn. or Activity	Species	Relative I uptake (%)
H_2O_2	5 µM	Ls	+24
	25 µM	Ls	+46
	50 µM	Ls	+12
	75 µM	Ls	–10
	250 µM	Ls	–33
	500 µM	Ls	–44
Ascorbate	2.5 mM	Ld	–86
	2.5 mM	Ls	–93
Cu-bi-salicylate	500 µM	Ld	–39
	5 mM	Ld	–86
Glutathione	500 µM	Ld	–47
	2.5 mM	Ls	–84
	5 mM	Ld	–94

Results

Iodine accumulation by Laminaria sporophytes. The iodine contents of sporophytes of *L. digitata* collected off Roscoff or cultured in tanks of running-sea water were compared (Fig. 1). The older the sporophytes the lower their average iodine contents, down to a level of 0.4% in 3- to 4-year old plants. In such adult sporophytes, younger, fast-growing tissue tended to have a lower iodine content than older parts. Plantlets (less than 15 cm in length), however, were very rich in iodine, 4.7% DW on average, equivalent to a concentration of 73 mM fresh tissue and an accumulation of 5 orders of magnitude relative to seawater. The total iodine content of *L. saccharina* protoplast preparations ranged from 0.33 to 0.74% in dry weight. Relative to cells in planta and considering that cell walls account for ca. 50% of dry weight in *Laminaria* (Mabeau and Kloareg 1987), protoplasts had lost from 50% to 80% of the intracellular iodine pool.

Iodine uptake by Laminaria sporophytes. The rate of iodine uptake by thallus discs punched out of marginal *L. saccharina* sporophyte tissue depended on the iodide concentration in the external medium. Michaelis-Menten kinetics were observed in the range of 2 µM $^{-1}$mM iodide concentrations (Fig. 2). A 500-fold increase in the iodide concentration in the surrounding medium (2 µM to 1 mM) raised the uptake rate from 2.8 nmol · (g tissue)$^{-1}$ · min^{-1} to 75.6 nmol · (g tissue)$^{-1}$ · min^{-1}, i.e., a 27-fold enhancement. At the highest concentration tested, about 1.5 µmol iodine was taken up per gram of tissue after 20 min. The apparent V_{max} and K_m values observed with tissue discs were 40.8–52.3 nmol · (g tissue)$^{-1}$ · min^{-1} and 26–47 µM iodide, respectively. The *L. digitata* tissue exhibited similar Michaelis-Menten kinetics, with apparent V_{max} and K_m values of 65.4 nmol · (g tissue)$^{-1}$ · min^{-1} and 420 µM iodide, respectively (not shown). Incubation in the presence of 4 µM dichlorophenyl-dimethylurea (DCMU), an inhibitor of the D1 protein of the photosynthetic electron transport chain, slightly enhanced the uptake of iodide. Preincubation for 15–60 min with the proton-gradient uncoupler DNP at concentrations as high as 1 mM did not have a significant inhibitory effect on iodine absorption.

Involvement of activated oxygen species (AOS) in iodine accumulation by Laminaria. In order to investigate the involvement of oxidative metabolism in iodine uptake, *Laminaria* sporophyte tissue discs were incubated with H_2O_2 over concentrations ranging from 5 µM to 500 µM. Enhancement of iodine uptake was highest (46%) upon addition of 25 µM H_2O_2 (Table 1). At H_2O_2 concentrations above ca. 60 µM, however, the uptake of radioactive iodine decreased significantly, down to one-half of control levels at concentrations around 1 mM. Preincubating *Laminaria* tissues for 1 h in the presence of various AOS scavengers such as glutathione, ascorbate or copper bi-salicylate also resulted in a reduction of iodine uptake, by 86–94% from

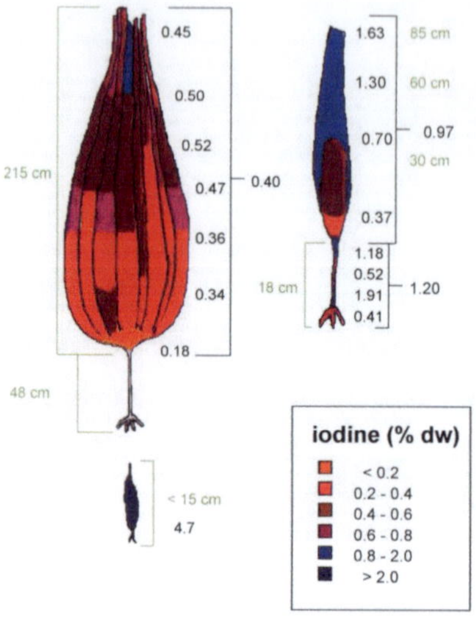

Fig. 1. Accumulation of iodine in sporophytes of *L. digitata*. Plants were collected at Enez Glas in Roscoff in the Spring of 1996. Iodine contents are displayed by different colours (key in the figure). Note that in 1-year-old plants (ca. 1 m long), iodine contents are lower in elongating areas above and below the meristematic zone, as well as in the holdfast. In young plantlets the iodine content, an average of 4.7% of dry weight, represents an accumulation of 150,000-fold relative to natural seawater

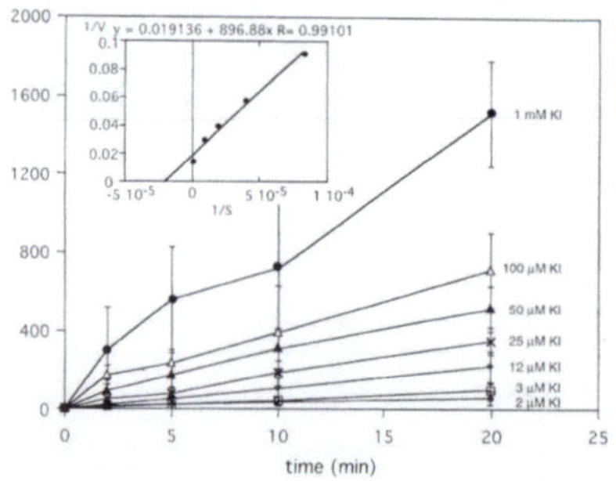

Fig. 2. Kinetics of iodine uptake by *L. saccharina* sporophyte discs in the presence of various concentrations of KI ($n = 5$). The amounts of iodine taken up are given on the ordinate in nanomoles per gram of tissue (FW). Error bars represent the standard deviation from the average value. The Lineweaver-Burk plot used for K_m and V_{max} determination is inserted

control levels (Table 1). The inhibitory effect of ascorbate was investigated over 4 orders of magnitude, between 1 µM and 10 mM. Inhibition of the uptake of iodine became significant at concentrations over 100 µM ascorbate and could be reversed upon removing the scavenger from the medium (not shown).

Laminaria digitata plantlets previously loaded with ^{125}I released iodine at an efflux rate of 1.87 nmol · (g tissue)$^{-1}$ · min^{-1}, corresponding to a loss of 0.15% per hour of the accumulated iodine (based on the average iodine content of such plantlets, see Fig. 1). In contrast, an immediate and marked release of iodine was

Table 2. Release of accumulated iodine upon oxidative stress. Young *L. digitata* plantlets pre-loaded with radioactive label were exposed to various peroxide concentrations. Most of the iodine was usually released within the first 15 min of the stress

H_2O_2 addition	% iodine released after 3 h
Control	0.06 ± 1.41
50 μM	4.16 ± 1.05
200 μM	5.55 ± 2.22
2 mM	7.81 ± 10.29

observable upon incubation in the presence of exogenous H_2O_2, from concentrations higher than 50 μM (Table 2). Incubation with 2 mM H_2O_2 resulted in an initial rate of efflux of 615.2 nmol · (g tissue)$^{-1}$ · min^{-1} and in the release of about 8% of the intracellular iodine reservoir within 10 min.

The uptake of iodine by protoplasts prepared from basal and marginal tissues of *L. saccharina* was compared with that of thallus discs, assuming that 5×10^7 protoplasts correspond to about 1 g of sporophyte tissue (Butler et al. 1989). In the presence of 2 μM iodide, the absorption rate by protoplasts (11 nmol · g^{-1} · min^{-1}) was only one-tenth of that measured in the tissue. However, upon addition of glucose and GOD as an extracellular H_2O_2-generating system, uptake of iodine was sustainably restored in protoplasts (Fig. 3). The rate of uptake sharply increased at a GOD concentration threshold around 0.01 U · ml^{-1} (corresponding to a supply of about 100 μM peroxide), reaching saturation at above 1 U · ml^{-1} (corresponding to steady-state concentrations of about 1 mM peroxide), and leading to the uptake of about 10% of the iodine available from the solution within 10 min. Control experiments using

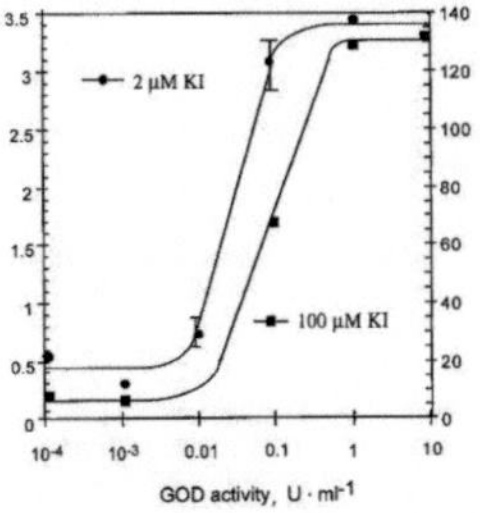

Fig. 3. Influence of the concentration of exogenous GOD on the uptake of iodine by isolated *L. saccharina* protoplasts. Protoplasts were incubated at two different KI concentrations in the presence of 100 mM glucose and of various GOD activities, ranging from 10^{-4}–10 U · ml^{-1}. The amounts of accumulated iodine are given on the ordinates in nanomoles taken up per $1.5 \cdot 10^8$ protoplasts within 20 min, for 2 μM KI (●, left-hand scale) and 100 μM KI (■, right-hand scale)

[^{14}C]-sorbitol, a polyalcohol known to be membrane-impermeable, indicated that under these conditions carry-over from the incubation medium was less than 0.6% after 1 h incubation.

Involvement of haloperoxidase in iodine uptake. Potential involvement of cell-wall haloperoxidases in iodine uptake (Vilter et al. 1983) was investigated using *L. digitata* gametophytic filaments and vanadate-dependent haloperoxidases purified from *Ascophyllum nodosum* (Fig. 4) and *L. digitata*. Constitutive levels of iodine uptake were much lower than in sporophytes, ca. 0.24 nmol · (g Fw)$^{-1}$ · min^{-1} in the presence of 10 μM KI (Fig. 4, treatment a). When provided with 10 μM exogenous H_2O_2 (Fig. 4, treatment b), gametophytes took up iodide at a rate of ca. 0.87 nmol · (g Fw)$^{-1}$ · min^{-1}. This rate was further increased 2-fold upon addition of haloperoxidase from *A. nodosum* (Fig. 4, treatment c). Even higher uptake rates, up to 2.7 nmol · (g FW)$^{-1}$ · min^{-1}, were obtained in the presence of 10 U · ml^{-1} A.n.I haloperoxidase when a second enzymatic system was added, consisting of 0.025 U · ml^{-1} GOD, 6.4 U · ml^{-1} catalase and 100 mM glucose and generating a constant level of 5 μM peroxide (Fig. 4, treatment e). Comparable results were obtained with the isoenzyme from *L. digitata* (data not shown).

Budget of iodine fluxes in Laminaria. In all uptake experiments with tissue discs, the removal of radioactive label from the medium (as determined by measurements in liquid aliquots) exceeded the activity actually taken up by *Laminaria* sporophytes. The deficit ranged between

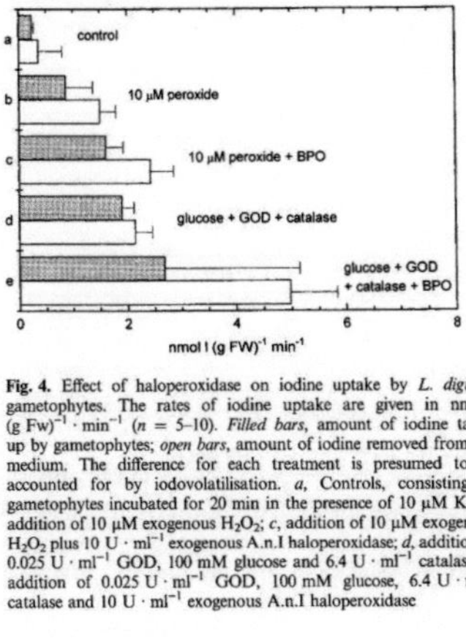

Fig. 4. Effect of haloperoxidase on iodine uptake by *L. digitata* gametophytes. The rates of iodine uptake are given in nmol · (g Fw)$^{-1}$ · min^{-1} (n = 5–10). *Filled bars*, amount of iodine taken up by gametophytes; *open bars*, amount of iodine removed from the medium. The difference for each treatment is presumed to be accounted for by iodovolatilisation. *a*, Controls, consisting of gametophytes incubated for 20 min in the presence of 10 μM KI; *b*, addition of 10 μM exogenous H_2O_2; *c*, addition of 10 μM exogenous H_2O_2 plus 10 U · ml^{-1} exogenous A.n.I haloperoxidase; *d*, addition of 0.025 U · ml^{-1} GOD, 100 mM glucose and 6.4 U · ml^{-1} catalase; *e*, addition of 0.025 U · ml^{-1} GOD, 100 mM glucose, 6.4 U · ml^{-1} catalase and 10 U · ml^{-1} exogenous A.n.I haloperoxidase

10 and 30% of the total iodine removed from the medium. Controls demonstrated that this deficit could not be accounted for by the washing procedure nor by a quenching of (γ-) rays by the setup for detecting radioactivity. A comparable deficit of radioactive iodine was also detected in oxidative efflux experiments with plantlets previously charged with radioactive iodine. Finally, when using *L. digitata* gametophytes in various uptake conditions, the amounts of label removed from the incubation medium were always significantly higher than the label recovered in the biomass (Fig. 4).

Discussion

Iodine uptake in Laminaria is a facilitated-diffusion mechanism. For an iodine concentration of 2 μM KI and upon incubation for 6 h, *L. saccharina* tissue took up iodine at the rate of 1 nmol · g^{-1} · min^{-1}. In a comparable experiment with tissue discs incubated in 1 μM KI the uptake rate was 0.7 nmol · g^{-1} · min^{-1} (Amat 1985). Consistent with the 17-fold enhancement observed by Amat (1985) upon increasing the external iodide concentration from 1 μM to 100 μM, raising the iodide concentration in seawater from 2 to 100 μM and to 1 mM resulted in a 12.8-fold and a 27-fold increase in the iodine uptake rate, respectively. Overall, the variations in the uptake of iodide by *L. saccharina* tissue with the external iodide concentration fit a Michaelis-Menten-like equation (Fig. 1). In agreement with the fact that Laminariales accumulate much more iodine than Fucales (Vilter et al. 1983), the apparent theoretical maximal speeds of iodine entry in kelp sporophyte tissue, ca. 45 nmol · g^{-1} · min^{-1} in *L. saccharina* and 65.4 nmol · g^{-1} · min^{-1} in *L. digitata*, are about 10 times higher than in *F. ceranoides* (V_{max} = 4.2–7.5 nmol · g^{-1} · min^{-1}; Baily and Kelly 1955).

One-hour preincubations with DNP at concentrations as high as 1 mM did not inhibit iodine uptake by *L. saccharina* tissues. Based on the reversible, inhibitory effect of a 20-h preincubation with 500 μM DNP, André (1965) suggested an active transport process for iodide in *L. digitata*. However, as this proton ionophore is likely to disturb the whole ATP-dependent metabolism, secondary effects caused by energy depletion cannot be excluded in such long-term experiments. Indeed, the uptake of iodine was not significantly affected until 6 h after the addition of DNP (André 1965). Similarly, no inhibitory effect was observed on iodine uptake by *L. saccharina* upon incubation in the presence of 10–20 μM carboxyatractiloside, an inhibitor of the mitochondrial ATP translocase which blocks the availability of ATP from respiration (Amat 1985). Altogether, it appears that iodine uptake follows Michaelis-Menten kinetics but that no direct ATP expenditure is required for the accumulation of iodine in *Laminaria*. According to the definition of Lobban et al. (1985), this process may therefore be thought of as a facilitated-diffusion mechanism, by which the rate of transport of iodide into *Laminaria* cells is increased, irrespective of its electrochemical potential gradient.

Apoplastic oxidative power regulates the influx and efflux of iodine. Hydrogen peroxide spontaneously reacts with iodide to form hypoiodous acid (1), virtually undissociated (pK = 10.64) at the pH (8.0) of seawater,

$$I^- + H_2O_2 \Longleftrightarrow HIO + OH^- \qquad (1)$$

$$HIO + I^- + H^+ \Longleftrightarrow I_2 + H_2O \qquad (2)$$

and itself in equilibrium (2) with molecular iodine in aqueous solutions (Truesdale 1995). If one or the other of the oxidised iodine species can freely cross the plasma membrane, one would expect iodine influx into *Laminaria* cells to be dependent on apoplastic oxidative power. The uptake of iodine by sporophyte tissue was indeed enhanced by low doses of exogenous peroxide. The dose-dependent inhibition of tracer accumulation by scavengers (Table 1) is likely to be due to competition with iodide in accepting electrons from AOS.

Only a very small constitutive net production of H_2O_2, ca. 0.5 nmol · (g Fw)$^{-1}$ · min^{-1}, is required in the apoplast to stoichiometrically account for the uptake of iodine from natural seawater. Assuming the apoplastic volume of *Laminaria* fronds at about 10% of the tissue fresh weight (Mabeau and Kloareg 1987), this basal production amounts to peroxide concentrations in the apoplast of ca. 5 μM. Consistently, *Laminaria* disks provided with exogenous H_2O_2 showed an enhanced uptake for concentrations of peroxide ranging from 5–50 μM. However, attempts to artificially further increase the apoplastic peroxide levels resulted in a marked inhibition of the influx of radioactive iodine (Table 1). As shown in Table 2, this is most likely due to a massive efflux of non-radioactive iodine followed by isotopic dilution of the external tracer. Consistently, methyl viologen, a redox-cycling compound generating massive quantities of AOS in plastids and other cellular compartments, inhibited iodine uptake by almost 30% when applied at the concentration of 20 μM.

In contrast to the blade tissue, protoplasts isolated from *L. saccharina* did not take up iodine at significant rates. In the presence of glucose/GOD, however, iodine influx into protoplasts correlated with the activity of the peroxide-generating system (Fig. 3). As carry-over from the surrounding medium was smaller by 2 orders of magnitude than actual accumulation of iodine, the influx of this element in *Laminaria* protoplasts supplied with H_2O_2 cannot be accounted for by membrane leakiness. However, whilst H_2O_2 concentrations higher than 50 μM were reducing iodine uptake in tissue, concentrations as high as 1 mM still enhanced net iodine uptake in protoplasts (Fig. 3). This probably reflects a significantly lower rate of iodine efflux from the isolated protoplasts, which had lost about half of the intracellular concentration of this element, probably as a result of the massive oxidative stress during enzymatic cell wall degradation (Benet et al. 1997).

As an uncharged and relatively stable molecule, H_2O_2 can easily cross membranes and reach locations remote from the site of its formation. In plant cells, AOS arise in plastids and mitochondria, from photorespiration or incomplete reduction of oxygen. Yet, in *Laminaria*, neither blocking the electron transport downstream of PSII with dichlorophenyl-dimethylurea nor diverting electrons from $NADP^+$ reduction into the Mehler reaction with methyl viologen markedly increased iodine uptake. Consistently, the influx of iodide is not reduced by darkness (e.g. Shaw 1959; Amat 1985). From these data, intracellular production of H_2O_2 does not appear to be a major source for the oxidation of apoplastic iodide. The AOS also can be generated extracellularly, by plasmalemmal NADPH oxidases (e.g. Baker and Orlandi 1995) or a variety of apoplastic oxidases and peroxidases (e.g., Mader et al. 1980; Dumas et al. 1993; Bolwell et al. 1995). In Laminariales, no plasmalemmal or apoplastic proteins involved in generation of H_2O_2 have been described yet. In particular, no "iodide oxidase" has been isolated so far. Such apoplastic enzymes, however, are likely to be lacking or inactivated in the isolated protoplasts. Restoration of iodine uptake in protoplasts by exogenous peroxide thus indicates that extracellular enzymes are the main providers of AOS for the accumulation of iodine in kelps.

Haloperoxidase catalyses iodine uptake in Laminaria. Murphy and O hEocha (1973b) have suggested that, in addition to a peroxide-generating system, iodide uptake in algae is mediated by an iodide peroxidase, which would catalyse the oxidation of iodide by peroxide. It is now well documented that cell walls of brown algae contain various isoforms of vanadium (V)-containing haloperoxidases (Vilter 1995), with activities as high as ca. $5\ U \cdot ml^{-1}$ in *Laminaria* apoplasm (Jordan et al. 1991). Haloperoxidases are known to catalyse the formation of hypohalides, according to reaction (3),

$$X^- + H_2O_2 \Longleftrightarrow HXO + OH^- \qquad (3)$$

where X^- stands for the halides Br^- or I^- (Vilter 1995). As reaction (3) is identical to equilibrium (1), one might expect that haloperoxidases are involved in iodine accumulation in brown algae (Jordan and Vilter 1991). Consistent with this hypothesis, in the presence of 6–$12\ U \cdot ml^{-1}$ haloperoxidase from either *L. digitata* or *A. nodosum* and a constant supply of $5\ \mu M\ H_2O_2$, *L. digitata* gametophytes took up iodine at rates markedly higher than controls and approaching those of sporophytes. As these H_2O_2 levels and peroxidase activities can be thought of as representative of the physiological apoplastic conditions in sporophytes, this finding indicates that haloperoxidase is involved in iodine absorption in kelps, by catalyzing iodide oxidation in the cell wall.

The mechanism and possible functions of iodine accumulation in Laminaria. Based on the above results, Fig. 5 illustrates our current view of the mechanism leading to iodide accumulation in *Laminaria*. Cell wall oxidases or membrane-bound enzymes with an extracellular domain, together with possible intracellular sources, generate a steady flow of H_2O_2 into the apoplast. Cell wall haloperoxidases catalyse the physiological oxidation of iodide into hypoiodous acid and molecular iodine. Oxidised iodine can then cross the plasma membrane. Some of the oxidised iodine, however, evolves into the air as molecular iodine or as volatile halocarbons (VHCs). As pointed out by Kelly and Baily (1951), such a iodovolatilisation is likely to account for the deficit of the radioactive tracer observed in both their and our experiments.

As lipophilic species, hypoiodous acid or molecular iodine diffuse freely through the plasmalemma and enter cells, where they are reduced to iodide or diverted to iodinating organic substrates. In this mechanism the

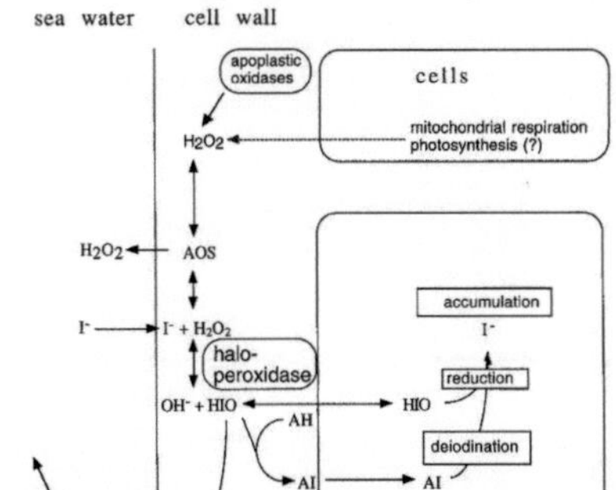

Fig. 5. Diagram showing the possible mechanism of iodine uptake in *Laminaria*. Apoplastic H_2O_2, mainly from extracellular sources is consumed for the oxidation of iodide into hypoiodous acid and molecular iodine, a reaction catalyzed by cell wall haloperoxidase. Oxidised iodide can then freely cross the plasmalemma, resulting in a facilitated diffusion influx of iodine. Haloperoxidase activity may also catalyze the iodination of organic compounds (*AI*) which would act as vectors of iodine into the cells. In steady-state conditions, a net influx of iodine is observed, leading to intracellular accumulation of iodine. Upon oxidative stress, this reservoir is mobilised and a rapid, massive efflux of iodine occurs. In both cases, oxidation of iodide results in the evolution of molecular iodine and of volatile halogenated compounds *VHCs* ("iodovolatilisation")

form of iodine transported across the plasma membrane is oxidised iodide whereas the intracellular stores of iodine mainly consist of reduced or organic iodine. Such a mechanism whereby iodine does not penetrate the cells against an electrochemical potential gradient is consistent with the facilitated diffusion kinetics reported above. This model extends the diagram put forward by Shaw (1959, 1960, 1962), with the addition of two refinements, i.e., the demonstration that oxidation of iodide takes place in the apoplast and that it is catalysed by haloperoxidase. At the thallus level, the distribution of accumulated iodine in *Laminaria* (Fig. 1) can be thought of as the result of two major fluxes, transport into the cells, mediated by both oxidases (still to be identified) and haloperoxidase, as well as long-distance transport (Amat and Srivastava 1985).

One puzzling question remains, the biological significance of the accumulation of iodine by kelps, up to levels as high as 73 mM in *L. digitata* plantlets. It has been proposed (e.g. Wever et al. 1991) that it serves the production of iodinated, antimicrobial molecules. A first defence line would then be the production of H_2O_2 in the apoplast. Such an oxidative burst would in turn trigger a rapid, massive mobilisation of iodine from the intracellular reservoir (this study), leading to the haloperoxidase-mediated formation of VHCs (Wever et al. 1991; Collén et al. 1994; and Fig. 4, this study) and participating in the protection against oxidative stress (Pedersén et al. 1996). The haloperoxidases of brown algae therefore appear to be key enzymes in iodine metabolism, both for its uptake and for its utilization in the production of VHCs.

Gouestlet eo ar penn labour-man da vezhinerien Vreizh, e koun an amzer gozh, da vare ar soud. Ce travail est dédié aux goëmoniers de Bretagne, en mémoire de l'époque de l'extraction de l'iode. This work is dedicated to Brittany's seaweed gatherers, as a tribute to the time of kelp burning.

The authors thank Prof. A. Pujin (University of Bourgogne, Dijon) and Prof. J. Guern (Institut des Sciences Végétales, Gif-sur-Yvette) for enriching discussions on oxidative metabolism in higher plants. Financial support from the FEDER Vb programme of the European Union to B.K. and from the Deutsche Forschungsgemeinschaft to H.V. is gratefully acknowledged. N.S. received a fellowship from the French Ministry of Foreign Affairs and F.K. was granted scholarships from Studienstiftung des Deutschen Volkes, Bonn, and Hüls AG-Stiftung, Marl. This is a contribution of GDR 1002 CNRS-IFREMER.

References

Amat MA (1985) Physiology of iodine uptake and distribution in *Laminaria saccharina* (Phaeophyta). Doctoral Thesis, Univ P et M Curie, Paris, France

Amat MA, Srivastava LM (1985) Translocation of iodine in *Laminaria saccharina* (Phaeophyta). J Phycol 21: 330–333

André S (1965) Sur la biochimie comparée du transport des iodures. C R Soc Biol Paris 159: 2327–2332

André S (1971) Destinée des iodures fixés chez les diverses algues marines et caractérisation des acides aminés iodés dans les hydrolysats. C R Soc Biol Paris 162: 2293–2298

Ar Gall E, Asensi A, Marie D, Kloareg B (1996) Parthenogenesis and apospory in the Laminariales: a flow cytometry analysis. Eur J Phycol 31: 369–380

Baily NA, Kelly S (1955) Iodine exchange in *Ascophyllum*. Biol Bull Woods Hole Mass 109: 13–20

Baker CJ, Orlandi EW (1995) Active oxygen in plant pathogenesis. Annu Rev Phytopathol 33: 299–321

Benet H, Ar Gall E, Asensi A, Kloareg B (1997) Protoplast regeneration from gametophytes and sporophytes from some species in the order Laminariales (Phaeophyceae). Protoplasma 199: 39–48

Black WAP (1948) The seasonal variation in chemical constitution of some of the sub-littoral seaweeds common to Scotland, *Laminaria cloustinii*, *L. digitata*, *L. saccharina*, and *Saccorhiza bulbosa*. J Soc Chem Ind (London) 67: 165–176

Bolwell GP, Butt VS, Davies DR, Zimmerlin A (1995) The origin of the oxidative burst in plants. Free Radicals Res 23: 517–532

Butler D, Otsgaard K, Boyen C, Evans LV, Jensen A, Kloareg B (1989) Isolation conditions for high yields of protoplasts from *Laminaria* species. J Exp Bot 40: 1237–1246

Butler A, Soedjak HS, Polne-Fuller M, Gibor A, Boyen C, Kloareg B (1990) Studies of vanadium-bromoperoxidase using surface and cortical protoplasts of *Macrocystis pyrifera* (Phaeophyta). J Phycol 26: 589–592

Collén J, Abrahamsson K, Ekdahl A, Pedersén M (1994) The involvement of hydrogen peroxide on the production of volatile halogenated compounds by *Meristiella gelidium* (Rhodophyta). Phytochemistry 36: 1197–1202

Dangeard P (1928) Sur le dégagement d'iode chez les algues marines. C R Acad Sciences Paris 186: 892–894

Dangeard P (1957) L'iode des algues, les ioduques, l'iodovolatilisation. Le Botaniste 41: 187–207

De Boer E, Tromp MGM, Plat H, Krenn GE, Wever R (1986) Vanadium (V) as an essential element for haloperoxidase activity in marine brown algae: purification and characterization of a vanadium (V)-containing bromoperoxidase from *Laminaria saccharina*. Biochim Biophys Acta 872: 104–115

Dumas B, Sailland A, Cheviet JP, Freyssinet G, Pallett K (1993) Identification of barley oxalate oxidase as a germin-like protein. C R Acad Sci Paris 316: 793–798

Glazener JA, Orlandi EW, Harmon GL, Baker CJ (1991) An improved method for monitoring active oxygen in bacteria-treated suspension cells using luminol-dependent chemiluminescence. Physiol Mol Plant Pathol 39: 123–133

Jordan P, Vilter H (1991) Extraction of proteins from material rich in anionic mucilages: partition and fractionation of vanadate-dependent bromoperoxidases from the brown algae *Laminaria digitata* and *L. saccharina* in aqueous polymer two-phase systems. Biochim Biophys Acta 1073: 98–106

Jordan P, Kloareg B, Vilter H (1991) Detection of vanadate-dependent bromoperoxidases in protoplasts from the brown algae *Laminaria digitata* and *L. saccharina*. J Plant Physiol 137: 520–524

Kelly S, Baily NA (1951) The uptake of radioactive iodine by *Ascophyllum*. Biol Bull 100: 188–190

Klemperer HG (1957) The accumulation of iodine by *Fucus ceranoides*. Biochem J 61: 381–389

Kylin H (1929) Über das Vorkommen von Jodiden, Bromiden und Jodidoxydasen bei Meeresalgen. Hoppe-Seyler's Z Physiol Chem 186: 50–84

Lobban CS, Harrison PJ, Duncan MJ (1985) Nutrients. In: Lobban CS, Harrison PJ, Duncan MJ (eds) The physiological ecology of seaweeds. Cambridge Univ Press, pp 75–110

Mabeau S, Kloareg B (1987) Isolation and analysis of the cell walls of brown algae: *Fucus spiralis*, *F. ceranoides*, *F. vesiculosus*, *F. serratus*, *Bifurcaria bifurcata* and *Laminaria digitata*. J Exp Bot 38: 1573–1580

Mader M, Ungemach J, Schloß P (1980) The role of peroxidase isoenzyme groups of *Nicotiana tabacum* in hydrogen peroxide formation. Planta 147: 467–470

Murphy MJ, O hEocha C (1973a) Peroxidase from the green alga *Enteromorpha linza*. Phytochemistry 12: 55–59

Murphy MJ, O hEocha C (1973b) Peroxidase activity in the brown alga *Laminaria digitata*. Phytochemistry 12: 2645–2648

Pedersén M, Collén J, Abrahamsson K, Ekdahl A (1996) Production of halocarbons from seaweeds: an oxidative stress reaction? Sci Mar 60 (Suppl. 1): 257–263
Roche J, Yagi Y (1952) Sur la fixation de l'iode radioactif par les algues et sur les constituants iodés des Laminaires. C R Soc Biol Paris 146: 642–645
Saenko GN, Kravtsova YY, Ivanenko VV, Sheludko SI (1978) Concentration of iodine and bromine by plants in the Seas of Japan and Okhotsk. Mar Biol 47: 243–250
Shaw TI (1959) The mechanism of iodine accumulation by the brown sea weed *Laminaria digitata*. The uptake of I^{131}. Proc R Soc Lond Ser B 150: 356–371
Shaw TI (1960) The mechanism of iodine accumulation by the brown sea weed *Laminaria digitata*. Respiration and iodide uptake. Proc R Soc Lond Ser B 152: 109–117
Shaw TI (1962) Halogens. In: Lewin RA (ed) Physiology and biochemistry of algae. Academic Press, New York, pp 247–253
Tong W, Chaikoff IL (1955) Metabolism of I^{131} by the marine alga, *Nereocystis luetkeana*. J Biol Chem 215: 473–484
Truesdale VW, Luther III GW, Canosa-Mas C (1995) Molecular iodine reduction in seawater: an improved rate equation considering organic compounds. Mar Chem 48: 143–150
Vilter H, Glombitza KW, Grawe A (1983) Peroxidases from Phaeophyceae I: extraction and detection of the peroxidases. Bot Mar 26: 331–340
Vilter H (1984) Peroxidases from Phaeophyceae: a vanadium (V)-dependent peroxidase from *Ascophyllum nodosum*. Phytochemistry 23: 1387–1390
Vilter H (1994) Extraction of proteins from sources containing tannins and anionic mucilages. Methods Enzymol 228: 665–672
Vilter H (1995) Vanadium-dependent haloperoxidases. In: Sigel H, Sigel A (eds) Vanadium and its role in life. Metal ions in biological systems 31. Dekker, New York, pp 325–362
Wever R, Tromp MGM, Krenn BE, Marjani A, Van Tol M (1991) Brominating activity of the seaweed *Ascophyllum nodosum*: impact on the biosphere. Environ Sci Technol 25: 446–449
Wong GTF (1977) The distribution of iodine in the upper layers of the equatorial Atlantic. Deep-Sea Res 24: 115–125
Wu CY, Lin G (1987) Progress in the genetics and breeding of economic seaweeds in China. Hydrobiologia 151: 57–61
Young EG, Langille WM (1958) The occurrence of inorganic elements in marine algae of the Atlantic provinces of Canada. Can J Bot 36: 301–310

8

Kelps produce inorganic iodine and iodocarbons of atmospheric significance for defense

Malin, G. §✳, Küpper, F.C. §†, Carpenter, L.J. ✳, Baker, A.R. ✳, Broadgate, W.J. ✳, Liss, P.S. ✳ & Kloareg, B. †

✳ School of Environmental Sciences, University of East Anglia,
Norwich, NR4 7TJ. U.K.

† UMR 1931 CNRS and Laboratoires Goëmar, Station Biologique de Roscoff,
F-29682 Roscoff, Bretagne, France.

§ These authors contributed equally to the work

The important role of volatile halocarbon compounds in ozone depletion has attracted public concern and stimulated research into their production and atmospheric chemistry[1,2]. In contrast with the dominant anthropogenic sources of many bromo- and chlorocarbons, volatile iodinated compounds are primarily naturally formed. The iodocarbons rapidly photolyse in the troposphere giving rise to iodine monooxide radicals (IO) which can interact with bromine and chlorine in rapid ozone destruction cycles[3,4]. The deposition of iodine to land via wet and dry precipitation, which is essential for mammalian health[5], completes the global iodine cycle[6]. The major source of atmospheric iodine is marine in origin[7], and a range of alkyl iodides are produced biogenically in seawater and emitted to the air[8,9]. Here we show for the first time that the oligoguluronate-triggered oxidative burst in the brown alga *Laminaria digitata* is accompanied by a rapid efflux of inorganic iodine and volatile iodocarbons. Oligoguluronates are a powerful new tool for investigating the biochemistry of biologically mediated oxidative stress in brown algae, and the biogeochemical consequences of iodine accumulation and release.

Seaweeds of the order Laminariales have long been known to sequester iodine from seawater. Iodine is an essential element for mammals and 30 % of the world's human population are considered at risk of iodine deficiency which can lead to goiter and brain damage[5,10]. Its role in photoautotrophs has not been elucidated, though it is commonly presumed to have an antimicrobial function[11]. Seaweeds also produce volatile halocarbons (e.g.[9,12-14]), but the formation mechanisms and functional significance of these compounds are not clear. Commonly they are assumed to be part of a general defense strategy against epiphytic microbes and herbivores[15], and/or result from haloperoxidation reactions which eliminate intracellular hydrogen peroxide (H_2O_2)[16,17].

All aerobic organisms suffer oxidative stress when the intracellular production of reactive oxygen species (ROS), superoxide (O_2^-), hydrogen peroxide (H_2O_2) and hydroxyl radicals ($^{\bullet}OH$), exceeds the cellular scavenging capacity. Nevertheless, the oxidative burst is a ubiquitous characteristic of animal and higher plant defense systems[18]. Recent studies show that recognition of cell wall fragments initiates a rapid oxidative burst in the red algae *Chondrus crispus* and *Gracilaria conferta*[19,20], and this response can eliminate epiphytic bacteria including agar degraders[21].

In previous studies we found that incubation of *L. digitata* with H_2O_2 mobilised intracellular iodine[22], and that oligoguluronate (GG), a purified cell wall degradation product, elicited a rapid oxidative burst via apoplastic H_2O_2 production[23]. Here we directly investigated iodine metabolism in *L. digitata* and found that GG elicitation of sporophytes previously labelled with ^{125}I caused a marked net efflux of iodine (Fig. 1). The rate of iodine efflux observed was 1.25 µmol g^{-1} FW h^{-1} which is far higher than the steady state iodine uptake rates of 2.8 nmol iodine g^{-1} FW h^{-1} reported before[22]. The chloride/anion channel inhibitors DIDS (4,4'-Diisothiocyanatestilbene-2,2'-disulfonic acid) and NPPB (5-nitro-2[3-phenylpropylamino-]benzoic acid) suppress the oxidative burst in higher plants[24], but they enhanced the GG-induced iodine efflux in *L. digitata* (data not shown) suggesting that a different transport mechanism operates for iodine in kelps than for chloride in higher plants. Comparing data for ^{125}I liberation with the activity left in thalli after an oxidative burst revealed an average deficit of recovery of 10 %, suggesting a considerable release of volatile iodine. Oxidative stress elicited with 2 mM peroxide caused a larger and faster ^{125}I efflux (Fig. 1), but a similar deficit was seen.

To obtain further insight into the speciation of the iodine released by *L. digitata*, iodide (I^-) and total reducible iodine (TRI) were determined using voltammetry[25]. The TRI fraction will include I^-, iodate (IO_3^-), and may also include hypoiodous acid (HOI), molecular iodine (I_2) and other unidentified oxidised iodine species. The difference between TRI and iodide concentration represents oxidised iodine (I_{ox}). *L. digitata* control thalli released iodine as I^- and I_{ox} in similar quantities (Fig 2a). GG-induced oxidative stress resulted in a rapid iodine efflux dominated by I^-, and after 3 hours I^- and I_{ox} had increased 26 and 10-fold relative to the control (Fig. 2b). Although this rapid efflux suggests a mobile intracellular pool of I^- and I_{ox}, fast chemical interconversions between iodine species mean that external speciation might not reflect the form of iodine released. Interestingly, *L. digitata* utilises I^- rather than IO_3^- and uptake involves haloperoxidases[22]. Interconversions between IO_3^- and I^- in seawater are commonly assumed to be biologically mediated[26], as the abiotic reduction and oxidation processes are thermodynamically unfavourable[27]. Our data for the release of I_{ox} identifies a first possible route for the biological oxidation of iodide.

Following the addition of GG to induce an oxidative burst, *L. digitata* released a range of volatile iodocarbons dominated by CH_2I_2, CH_3I, $CHBr_2I$ and C_2H_5I into the incubation seawater (Fig. 3). In contrast, the production of bromocarbons generally decreased following the oxidative burst (Fig. 3). The total measured volatile iodocarbon release was 3 orders of magnitude lower than the efflux of inorganic iodine, and hence represents a small part of the total iodine efflux. This elevated release of iodocarbons following an oxidative burst is in agreement with observations of high atmospheric iodocarbon and IO levels in the coastal zone at low tide[28,29]. *L. digitata* was also incubated with H_2O_2 which caused a general increase in most of the iodocarbons, but markedly elevated levels of some bromocarbons (Fig. 3). Pedersen *et al.*[17] found that adding H_2O_2 to *Meristiella gelidium* increased production of bromo- and chlorocompounds, but that iodocarbon levels were unaltered. Although oxidative stress usually implies elevated extracellular H_2O_2 levels, exogenous H_2O_2 addition appears to be not very realistic in simulating physiological conditions of oxidative stress. Endogenous elicitors of the oxidative burst provide a more realistic tool for initiating cellular ROS formation.

In the intertidal environment seaweeds are susceptible to oxidative stress due to high light, CO_2 and nutrient limitation, desiccation as well as temperature and salinity changes. Furthermore, all seaweeds are susceptible to grazers and pathogens. Rapid recognition of cell wall fragments, coupled with induction of a defensive oxidative burst which released cytotoxic ROS and inhibitory inorganic and organic iodine compounds could confer selective value. The higher plant oxidative burst can also activate secondary defense responses including production of low molecular weight antimicrobial compounds[30], and some molecules of this type function as infochemicals in plant-to-plant, plant-to-predator and plant-herbivore mutualistic signalling[31]. The halocarbons produced by seaweeds have previously only been considered waste products of H_2O_2 scavenging by peroxidase enzymes[16,17]. The data presented here shows that consideration should be given to direct physiological roles for these iodocarbons. In particular, the antimicrobial properties of iodine are well known and new evidence shows that iodo- and bromocarbons can kill red algal pathogens (Bouarab & Potin, unpublished). Furthermore, iodide may have antioxidant function in humans[32], so the rapid efflux of inorganic iodine in *L. digitata* might also have an antioxidant role. We suggest that these metabolic functions reflect the need for iodine accumulation in the Laminariales.

Current radiation levels can cause oxidative stress in seaweeds[33], with the early life stages of *Laminaria* species being most susceptible at low tides[34]. Climate change is forcing increased temperatures, desiccation and UV-B fluxes in many parts of the world. Ignoring possible changes in biogeography, this would make seaweeds more susceptible to oxidative stress, which could lead to enhanced coastal emissions of volatile halogenated compounds and changes in the biogeochemical cycle of iodine.

Methods

Plant material.

Juvenile *L. digitata* (L.) LAMOUR. sporophytes were collected by diving near Roscoff (Brittany, France) and the Mace Head Atmospheric Research Station (County Galway, Eire). When appropriate, field-collected *L. digitata* thalli (2 to15 cm) were cultivated in aerated running seawater at ambient temperature and with a 16:8 h light (50-60 μmol photons m^{-2} sec^{-1}) dark cycle. Field collected algae were kept in fresh seawater for a maximum of 48 h and then incubated in filtered seawater in 2 litre glass vessels with ground-glass stoppered ports. Seawater only controls were included in all experiments. During field experiments the vessels were incubated outdoors in a large tank of seawater to reduce temperature changes.

Elicitor Preparation.

Oligoguluronate (GG) was prepared from *L. hyperborea* stipe sodium alginate as described previously[23]. In elicitation experiments 150 μg mL^{-1} GG or 2 mM H_2O_2 were used.

Radiotracer Studies.

To assess the effect of GG elicitation on iodine uptake *L. digitata* thalli (50 mg FW mL^{-1}) were incubated in small volumes of seawater containing 10 μM KI and 126 Bq ml^{-1} ^{125}I (as iodide; Amersham). GG and H_2O_2 induced iodine efflux was quantified using plantlets precharged over 48 h with 23 Bq ml^{-1} ^{125}I. See Küpper *et al.*[22] for further details.

Inorganic iodine speciation.

Seawater aliquots (12 ml) were removed time-wise from the incubation vessels, frozen at -20 °C and transported on dry-ice. Samples were diluted 10-200 fold with Milli-Q water, iodide (I^-) was determined using square wave cathodic stripping voltammetry and TRI was analysed after ascorbic acid reduction of oxidised iodine species[25]. Speciation could not be assessed after H_2O_2 elicitation due to iodide oxidation.

Trace gas analysis.

Aqueous phase halocarbon compounds were preconcentrated from 20 mL samples using a cryogenic purge-and-trap system and identified and quantified with a Hewlett Packard 6890/5973 gas chromatograph/mass spectrometer[9].

Acknowledgements

Funding was provided by the U.K. Natural Environment Research Council and fellowships from Studienstiftung des Deutschen Volkes (Bonn) and the MAST III programme of the European Commission. We thank the EC PARFORCE project (New Particle Formation and Fate in the Coastal Environment) for planning the Mace Head campaign, B. Larsen for preparing sodium alginate and Alain Heyraud, Sylvie Rousvoal, J.-M. Legendre and Gareth Lee for technical support.

References

1. Houghton, J. T. *et al.* (eds.) *Climate Change 1995. The Science of Climate Change.* (Cambridge University Press, Cambridge, 1996).
2. Ennis, C. A. (ed.) *Scientific assessment of ozone depletion: 1998* (World Meteorological Organization, Geneva, 1999).
3. Chameides, W. L. & Davis, D. D. Iodine: its possible role in tropospheric photochemistry. *Journal of Geophysical Research* 85, 7383-7398 (1980).
4. Solomon, S., Garcia, R. R. & Ravishankara, A. R. On the role of iodine in ozone depletion. *Journal of Geophysical Research* 99, 20491-20499 (1994).
5. Dobson, J. E. The Iodine Factor in Health and Evolution. *Geographical Review* 88, 1-28 (1998).
6. Wong, G. T. F. The Marine Geochemistry of Iodine. *Reviews in Aquatic Sciences* 4, 45-73 (1991).
7. Liss, P. S. & Slater, P. G. Flux of gases across the air-sea interface. *Nature* 247, 181-184 (1974).

8. Murphy, C. D., Moore, R. M. & White, R. L. An isotopic labeling method for determining production of volatile organohalogens by marine microalgae. *Limnology & Oceanography* 45, 1868–1871 (2000).

9. Carpenter, L. J., Malin, G., Küpper, F. C. & Liss, P. S. Novel iodine-containing trihalomethanes and other reactive halocarbons in the Coastal East Atlantic. *Global Biogeochemical Cycles* 14, 1191-1204 (2000).

10. Hetzel, B. S. & Dunn, J. T. The Iodine Deficiency Disorders: Their Nature and Prevention. *Annual Review of Nutrition* 9, 21-38 (1989).

11. Fenical, W. in *Marine Organic Chemistry* (eds. Duutsma, E. K. & Dawson, R.) 375 (Elsevier, New York, 1981).

12. Manley, S. L. & Dastoor, M. N. Methyl iodide (CH_3I) production by kelp and associated microbes. *Marine Biology* 98, 477-482 (1988).

13. Nightingale, P. D., Malin, G. & Liss, P. S. Production of chloroform and other low molecular weight halocarbons by some species of macroalgae. *Limnology and Oceanography* 40, 680-689 (1995).

14. Giese, B., Laturnus, F., Adams, F. C. & Wiencke, C. Release of volatile iodinated C-1-C-4 hydrocarbons by marine macroalgae from various climate zones. *Environmental Science & Technology* 33, 2432-2439 (1999).

15. Wever, R., Tromp, M. G. M., Krenn, B. E., Mariani, A. & van Tol, M. Brominating activity of the seaweed *Ascophyllum nodosum*: impact on the biosphere. *Environmental Science and Technology* 25, 446-449 (1991).

16. Collen, J. & Pedersen, M. A. Stress-Induced Oxidative Burst in *Eucheuma platycladum* (Rhodophyta). *Physiologia Plantarum* 92, 417-422 (1994).

17. Pedersen, M., Collen, J., Abrahamsson, K. & Ekdahl, A. Production of halocarbons from seaweeds: An oxidative stress reaction? *Scientia Marina* 60, 257-263 (1996).

18. Groom, Q. J., Torres, M.A., Fordham-Skelton, A.O., Hammond-Kosack, K.E., Robinson, N.J., Jones, J.D.G. rbohA, a rice homologue of the mammalian gp91phox respiratory burst oxidase gene 10(3),. *The Plant Journal* 10, 515-522 (1996).

19. Bouarab, K., Potin, P., Correa, J. & Kloareg, B. Sulfated oligosaccharides mediate the interaction between a marine red alga and its green algal pathogenic endophyte. *Plant Cell* 11, 1635-1650 (1999).

20. Weinberger, F., Freidlander, M. & Hoppe, H.-G. Oligoagars elicit a physiological response in *Gracilaria conferta* (Rhodophyta). *Journal of Phycology* 35, 747-755 (1999).

21. Weinberger, F. & Friedlander, M. Endogenous and exogenous elicitors of a hypersensitive response in *Gracilaria conferta* (Rhodophyta). *Journal of Applied Phycology* 12, 139-145 (2000).

22. Küpper, F.C., Schweigert, N., ArGall, E., Legendre, J.M., Vilter, H., Kloareg, B.,. Iodine uptake in Laminariales involves extracellular, haloperoxidase-mediated oxidation of iodide. *Planta* 2, 163-171 (1998). (Chapter 7 of this thesis)

23. Küpper, F. C., Kloareg, B., Guern, J. & Potin, P. Oligoalginates elicit an oxidative burst in the brown algal kelp *Laminaria digitata*. *Plant Physiology* 125, 278-291 (2001). (Chapter 6 of this thesis)

24. Jabs, T., Tschöpe, M., Colling, C., Hahlbrock, K. & Scheel, D. Elicitor-stimulated ion fluxes and O_2^- from the oxidative burst are essential components in triggering defense gene activation and phytoalexin

synthesis in parsley. 94:. *Proceedings of the National Academy of Sciences of the U.S.A.* 94, 4800-4805 (1997).

25. Campos, M. L. A. M. New approach to evaluating dissolved iodine speciation in natural waters using cathodic stripping voltammetry and a storage study for preserving iodine species. *Marine Chemistry* 57, 107-117 (1997).
26. Wong, G. T. F. Coupling iodine speciation to primary, regenerated or "new" production: a re-evaluation. *Deep-Sea Research Part I* 48, 1459-1476 (2001).
27. Luther, G. W. I., Wu, J. F. & Cullen, J. B. Redox chemistry of iodine in seawater - frontier molecular orbital theory considerations. *Advances in Chemistry* 244, 135-155 (1995).
28. Carpenter, L. J. *et al.* Short-lived alkyl iodides and bromides at Mace Head, Ireland: links to biogenic sources and halogen oxide production. *Journal of Geophysical Research* 104, 1679-1689 (1999).
29. Alicke, B., Hebestreit, K., Stutz, J. & Platt, U. Iodine oxide in the marine boundary layer. *Nature* 397, 572-573 (1999).
30. Ebel, J. Oligoglucoside elicitor-mediated activation of plant defense. *Bioessays* 20, 231-237 (1998).
31. Kessler, A. & Baldwin, I. T. Defensive Function of Herbivore-Induced Plant Volatile Emissions in Nature. *Science* 291, 2141-2144 (2001).
32. Venturi, S. *et al.* Role of iodine in evolution and carcinogenesis of thyroid, breast and stomach. *Advances in Clinical Pathology* 4, 11-17 (2000).
33. Franklin, L. A. & Forster, R. M. The changing irradiance environment: consequences for marine macrophyte physiology, productivity and ecology. *European Journal of Phycology* 32, 207-232 (1997).
34. Dring, M. J., Makarov, V., Schoshina, E., Lorenz, M. & Luning, K. Influence of ultraviolet radiation on the chlorophyll fluorescence and growth in different life-history stages of three species of *Laminaria* (Phaeophyta). *Marine Biology* 126, 183-191 (1996).

Figures

Fig. 1. Efflux of iodine after challenge of ^{125}I-loaded *Laminaria digitata* with 150 µg mL^{-1} oligoguluronate (—●—) or 2 mM (—■—) H_2O_2 added at time 0 mins. The control plantlets were incubated in medium without additions (---○---).

Fig. 2. Iodide (—△—) and total reducible iodine (—▲—) efflux from a. untreated control *Laminaria digitata* and b. *Laminaria digitata* challenged with 150 µg mL^{-1} oligoguluronate at time 0 mins.

Fig. 3. Release of volatile low molecular weight halocarbons from *Laminaria digitata* plantlets **A.** control (---○---) and following elicitation of an oxidative burst by addition of 150 µg mL^{-1} oligoguluronate at time 0 mins (—●—), and **B.** control (—□—) and after addition of 2 mM H_2O_2 at time 0 mins (—■—).

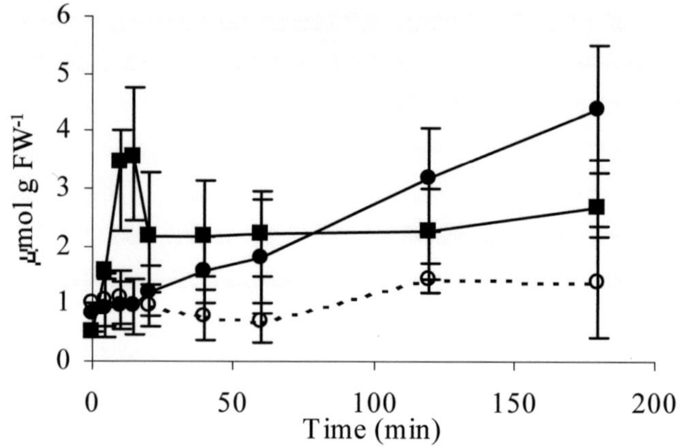

Figure 1

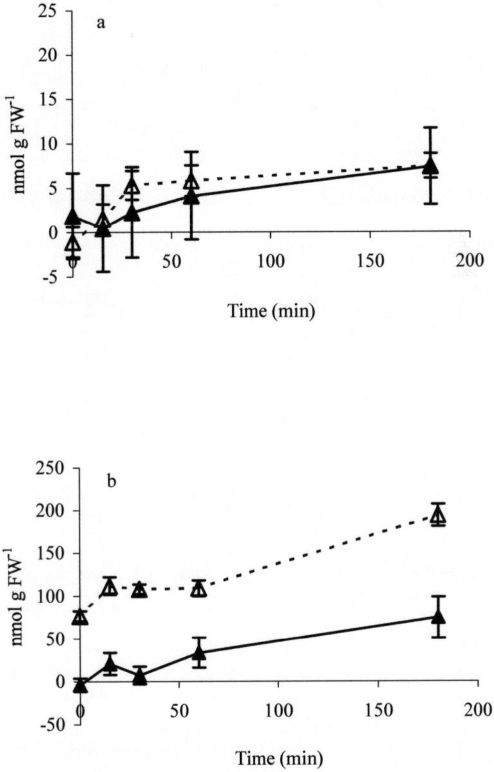

Figure 2

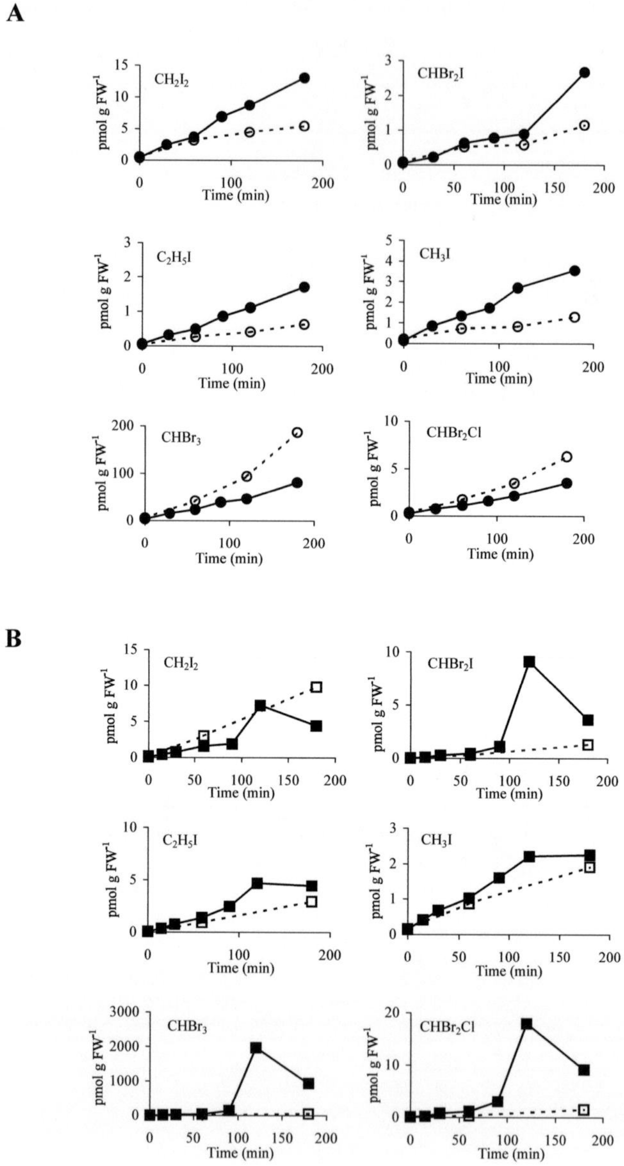
A
CH2I2
CHBr2I
C2H5I
CH3I
CHBr3
CHBr2Cl
B
CH2I2
CHBr2I
C2H5I
CH3I
CHBr3
CHBr2Cl
pmol g FW-1
Time (min)

Figure 3

9

Oligoalginate recognition and oxidative burst play a key role in natural and induced resistance of the sporophytes of Laminariales

Frithjof C. Küpper[1), 2)], Dieter G. Müller[2)], Akira F. Peters[3)], Bernard Kloareg[1)], Philippe Potin[1)*]

[1)]Station Biologique, UMR 1931, CNRS-Laboratoires Goëmar, BP 74, F-29682 Roscoff, Brittany, France
[2)]University of Konstanz, Dept. of Biology, D-78457 Konstanz, Germany
[3)]Institute for Marine Research, Düsternbrooker Weg 20, D-24105 Kiel, Germany
*corresponding author

Running title

Oligoalginates induce resistance in kelps

Abstract

45 species of brown algae (Phaeophyceae) were surveyed for their capacity to respond by an oxidative burst to challenging with alginate oligosaccharides. Intertidal frondose brown algae (Fucales) were shown to constitutively release high quantities of peroxide. The capacity to recognize oligoguluronates and to react with an oxidative burst was confined to alginate-rich taxa with complex thallus morphology, epitomized by the sporophytes of Laminariales. When kelp sporophytes were impaired in their capacity to perform an oxidative burst by the NADP(H) oxidase inhibitor diphenylene iodonium, they were readily degraded by their bacterial epiflora. Thus, in these algae, the oxidative response is an essential element of natural resistance. We also report on the establishment of a well-defined experimental system for investigations on kelp immunity, with Laminaria digitata as the host and its phaeophycean endophyte, Laminariocolax tomentosoides, as the pathogen. We found that an alginate-triggered oxidative burst induces resistance in Laminaria digitata against infection. From these findings we infer that oligoalginate signals seem to be important cues in the interaction between Laminarialean kelps and potential pathogens.

Key words

Marine algae - kelps - Laminaria – Laminariocolax – Macrocystis - *oligoalginates - oxidative burst - natural and induced resistance*

Abbreviations

DPI = diphenylene iodonium, FW = fresh weight, GG = oligoguluronates, ROS = reactive oxygen species

Introduction

All sessile plants, from marine benthic algae to terrestrial seed plants, are subject to attacks by grazers and pathogens. In terrestrial ecosystems, Olff and Ritchie (1998) have underlined the effects of grazing on plant diversity. Reciprocally, it was shown that changes in the levels of plant chemical defenses can govern the population dynamics of common herbivores (Koricheva et al. 1998) and the functioning (e.g. nutrient dynamics) of entire ecosystems (Penuelas & Estiarte 1998, Northup et al. 1998). Terrestrial plants have also evolved a variety of defenses against potentially harmful micro-organisms. Reminiscent of the innate immunity in animal lineages, plant immunity is based on the perception of cell-cell recognition signals referred to as elicitors (Scheel and Parker 1990, Boller 1995, Fry et al. 1993, Boudart et al. 1998). Endogenous elicitors consist of degradation products from plant surface components, such as pectin oligosaccharides, which are released from the host cell walls by the pectinolytic enzymes of pathogens (Boudart et al. 1998, Côté et al. 1998) or during herbivore damage (Bergey et al. 1999). Exogenous elicitors refer to pathogen signals released during the interaction, such as chitin or glucan fragments from fungal cell wall polysaccharides (Viard et al. 1994).

Both endogenous and exogenous elicitors are known to trigger a variety of defense responses. These include the activation of primary and secondary metabolism (Somssich and Halbrock 1998), the production of anti-microbial low-molecular weight secondary metabolites referred to as phytoalexins (Ebel et al. 1995, Apostol et al. 1987), the synthesis of proteins with anti-microbial activity known as pathogenesis-related (PR) proteins (Hammond-Kossack & Jones 1996, Fritig et al. 1998), and cross-linking of cell wall proteins (Otte & Barz 1996, Brisson et al. 1994), which eventually results in the control or the death of pathogenic fungi or bacteria (e.g., Ricci et al. 1989, Nürnberger et al. 1994). Elicitor recognition, at the surface or in the cytosol of host plant cells (Ebel and Mithöfer 1998), is followed whithin minutes by a so-called oxidative burst, which consists in the accumulation of reactive oxygen species (ROS), such as superoxide ions, hydrogen peroxide, or hydroxyl radicals, via the activation of plasma membrane-associated NAD(P)H oxidases or apoplastic peroxidases. ROS are intrinsically toxic, serve as second messengers and, as peroxidation substrates, they participate in cell wall cross-linking (e.g. Scheel 1998).

In marine macroalgae, competition and predation have been thought of as the major factors determining the structure of natural populations and communities (e.g. Carpenter 1990, Norton et al. 1990). Interactions with pathogens were considered too rare to have any significant impact on seaweed populations. Therefore, the importance of defense reactions against pathogens for the understanding of the functioning of marine ecosystems has been somewhat overlooked, and deciphering host-pathogen interactions in algae is still in its pioneering phase (Correa 1997, Potin et al. 1999).

Only a few marine pathosystems have been studied in some detail, mainly in the Rhodophytes. Weinberger et al. (1999) described an oxidative burst triggered by oligoagars in *Gracilaria conferta*, showing that it induces the elimination of agar-degrading epiphytic bacteria from the thallus surface (Weinberger & Friedlander 2000). Bouarab et al. (1999) demonstrated the role of carrageenan sulfation patterns in regulating the virulence of *Acrochaete operculata*, a pathogen of *Chondrus crispus*. Similar observations were made in the case of a sulfated agaroid known as porphyran, which was shown to enhance appressorium formation in *Pythium porphyrae*, the agent of the red-rot disease in *Porphyra yezoensis* mariculture (Uppalapati & Fujita 2000).

With respect to brown algae, we showed recently that oligomeric degradation products of alginate, the main matricial cell wall polysaccharide in this lineage, elicit an oxidative burst in the sporophytes of the kelp *Laminaria digitata* and trigger a cascade of signalling events (Küpper et al. 2001). Alginate is a linear polysaccharide made of homopolymeric blocks of β-1,4-D-mannuronic acid (oligomannuronates) or α-1,4-L-guluronic acid (oligoguluronates), and blocks with mixtures of both monomers (Haug et al. 1974). At micromolar concentrations or less, the homo-oligomeric fragments composed of guluronate residues elicited a transient, yet strong oxidative response, whereas polymannuronate blocks were inactive in *L. digitata* (Küpper et al. 2001).

Laminariales (kelps) are colonized by various morphologically distinguishable bacteria (Pellegrini & Pellegrini 1982, Corre et al. 1989), including alginate-degrading strains such as *Pseudomonas alginovora* (Boyen et al. 1990a). In Japanese mariculture systems of *Laminaria japonica* the alginolytic bacterium *Pseudoalteromonas elyakovii* was identified as the causative agent of the wounded-spot disease (Sawabe et al. 2000). In natural populations kelp sporophytes are also frequently infested by phaeophycean endophytes, resulting in severe symptoms such as warts, crippled thalli and twisted stipes and blades (Apt 1988, Kawai and Tokuyama 1995, Peters & Schaffelke 1996, Peters and Ellertsdóttir 1996, Ellertsdóttir and Peters 1997, Burkhardt and Peters 1998). Field observations (A.F. Peters, unpublished) suggest that mortality of such endophyte-infested adults is increased, mainly because stipes of infected individuals cannot resist the increased drag. In mature sporophytes, the presence of endophytes inhibits sorus formation (Peters & Schaffelke 1996, Lüning et al. 2000).

Altogether, pathogenic microorganisms are likely to be an important parameter in the fitness of these brown algae, either at the individual or the population levels.

Based on the biochemistry of host-pathogen interactions in terrestrial plants and on the above-mentioned observations in the Rhodophytes, one might thus expect that recognition of alginate oligosaccharides followed by an oxidative response is a key point in the defense strategies of brown algae against microbial epiphytes or pathogens. However, it is not known whether brown macroalgae other than the morphologically relatively complex and perennial *Laminaria digitata* also recognize this elicitor. We, therefore, have surveyed a number of brown algal groups, including species and life-history stages with a simple morphology, for their ability to sense the presence of oligoguluronate fragments, as shown by the release of ROS in their surrounding medium. We found that oligoguluronate sensing is restricted to the Phaeophyceae with a complex thallus morphology, such as the sporophytes of Laminariales.

The potential role of the oxidative burst in the natural resistance of kelps was then further investigated by loss-of-function experiments, using a suicide inhibitor of NAD(P)H oxidase, diphenylene iodonium. We demonstrate that inhibiting the capacity to perform an oxidative burst abolishes the natural resistance of kelp sporophytes towards their bacterial epiflora. Finally, we report on controlled inoculations using a convenient host-pathogen system to assess the biochemical basis of immunity in brown algae, using sporophytes of *Macrocystis pyrifera* and *L. digitata* and their natural endophytic pathogens of the genus *Laminariocolax*. In gain-of-function experiments by which alginate elicitors were used to trigger an oxidative burst in *L. digitata,* kelp sporophytes were protected from infections by *L. tomentosoides.* These findings indicate that recognition of alginate oligosaccharides and the related oxidative response play a role in the control of microorganisms by kelp sporophytes.

Methods

Algal and microbial materials

We used field specimens of various brown algal species collected from the seashore near Roscoff (Brittany, France) and Kiel (Baltic Sea, Germany). Additional species were grown as unialgal laboratory cultures originating from various localities (Table 1). Axenic sporophytes of *Macrocystis pyrifera* were obtained from a pair of axenic gametophytes from Chile (Magellan Strait). The brown algal endophytes, *Laminariocolax tomentosoides* (FARLOW) KYLIN (LtoH) and *L. macrocystis* (PETERS) PETERS (Lmac LM2) have been described previously (Burkhardt & Peters 1998, Peters 1991). Algae were grown in Provasoli ES medium (Starr & Zeikus 1987) prepared with natural sea water. Cultures, illuminated with daylight-type fluorescent lamps at an irradiance of 9 µmol photons m^{-2} s^{-1} for 10 h per day, were maintained at 10 ±1°C in 10 mL petri dishes and transferred to fresh medium at one- or two-week intervals.

The oomycete *Eurychasma dicksonii* (WRIGHT) MAGNUS was available in a unialgal culture in its host *Pylaiella littoralis* (L.) KJELLM., as described previously (Müller et al. 1999). The epiphytic bacterium *Pseudomonas alginovora* isolated from *Laminaria digitata* (Corre et al. 1989) was grown in ZoBell medium (ZoBell and Upham 1944) as previously described (Boyen *et al.* 1990a).

Measurements of hydrogen peroxide levels

Experiments were carried out in a 12 °C temperature-controlled room, except for warm-water species (*Dictyota phlyctaenodes*, *Zosterocarpus oedogoniae*), which were tested at 20 °C. Hydrogen peroxide concentrations were measured using the luminol chemiluminescence method with a LUMAT LB 9507 luminometer as described before (Küpper et al. 2001). Both the total amounts of hydrogen peroxide in the medium around plants and the kinetics of H_2O_2 release by brown algal cultures (50 – 500 mg blotted fresh weight in 10 mL of 0.22 µm-filtered seawater) either in the presence or in the absence of oligoguluronates (150 µg. mL^{-1}) were monitored. Negative results with oligoguluronates were validated by treatment with the K^+ ionophore valinomycin (5 µM). This potassium carrier had previously been shown to activate the burst-producing machinery downstream of GG recognition (Küpper et al. 2001). 400 µl aliquots (out of 10 mL samples) were taken per measurement.

Cell-free extracts of *Laminariocolax tomentosoides* were prepared as follows. Unialgal cultures were blotted dry on filter paper, weighed, frozen in liquid nitrogen and ground in an Eppendorf tube with a micropestle and Fontainebleau sand. To a fresh weight of ca. 20 mg algae, 300 µl extraction buffer (50 mM TRIS HCl, 50 mM NaCl, 10 mM $MgCl_2$, 1 mM PMSF; all from SIGMA) were added. The samples were vortexed for 1 min, incubated for 5 min (4 °C) and centrifuged for 5 min at 10000 *g*. The supernatant was recovered and stored at –20 °C until use. 50 µl of this extract were added to the *L. digitata* sporophytes either 1h prior to or simultaneously with the addition of alginate oligosaccharides. A similar procedure was used to prepare cell-free extracts from *Pylaiella littoralis* infected or not by *Eurychasma dicksonii.*

Challenge with epiphytic bacteria and with endophytes

The involvement of oxidative burst in the natural resistance of kelps to their natural epiflora was assessed as follows. Axenic sporophytes of *Macrocystis pyrifera* 2-5 mm in size were incubated for 10 min in the presence of the NADPH oxidase inhibitor diphenylene iodonium (1, 2.5, 5 and 10 μM from a 1 mM stock solution in DMSO; from SIGMA) in groups of 5 individuals, then transferred to fresh medium. In some cases, plants were elicited with oligoguluronates at 150 μg. mL^{-1} for 1 h before the DPI treatment. The thalli were then inoculated with *Pseudomonas alginovora* (4.5 10^8 colony forming units per 3.1 mL well), and inspected daily for the occurrence of lesions. Alternatively, the natural bacterial epiflora of *M. pyrifera* was re-established by adding a droplet of culture medium containing *Macrocystis*-associated bacteria from Chile.

To address the potential role of oligoalginate signals in the induced resistance of kelps, *L. digitata* sporophytes were elicited with oligoguluronates (150 μg. mL^{-1}) for 1 h, transferred to 50 ml fresh culture medium and inoculated at various intervals (1-14 days) after elicitation by *L. tomentosoides* zoospores, as follows. 500 μl of a culture of sporulating filamentous endophytes were added to each dish of *Laminaria*. After co-culture for 3 days, the *Laminaria* sporophytes were transferred to fresh medium, which was then renewed weekly, and they were examined at three-day intervals until 3 weeks after the inoculation. Since a numerical scoring of infections was not feasible, the state of the plants was judged by an unbiased, independent observer, according to the following relative scale: 0) massive infections, all plants dead; 1) strong infections, at least one plant dead; 2) strong infections and necroses in several plants; 3) strong infections or necroses in at least one plant; other plants of the same batch in good shape; 4) few infections or necroses: overall plants in good shape; 5) infections and / or necroses only very marginal, 6) no symptoms of pathogen infection nor unusual necroses. After the experiments, all plants were fixed and mounted on paper. They were deposited in the herbarium of the Marine Station, Roscoff.

Results

The oxidative response in brown algae upon treatment with oligoguluronates

45 species belonging to 11 orders of brown algae were monitored for the levels of hydrogen peroxide in their surrounding medium, in the absence or in the presence of alginate oligosaccharides (Table 1). In members of the order Laminariales, steady-state peroxide concentrations in the oligoalginate-free culture dishes were usually below 1 µM for the gametophytes, and below 500 nM for the sporophytes. In *L. digitata* and *L. ochroleuca* sporophytes, the H_2O_2 concentrations were typically in the range of 200 nM. Only in the culture of female *Chorda filum* gametophytes higher H_2O_2 concentrations were measured (Table 1).

When challenged with oligoguluronates, culture-raised, young sporophytes of *Chorda filum*, sporophytes of all the *Laminaria* species tested, as well as those of *Macrocystis pyrifera* (Fig. 1A) and of *Lessonia nigrescens*, reacted with a rapid, strong oxidative burst. Rates of H_2O_2 accumulation were typically as high as 0.3 – 0.5 µmol / g fresh weight (FW) for cm-long parenchymatic thalli within 15 min of the addition of alginate oligosaccharides (due to the higher surface / volume ratio, this value could be up to an order of magnitude higher for smaller thalli). In the case of *L. digitata*, *L. ochroleuca* and *L. pallida*, accumulation of H_2O_2 was very transient, with complete scavenging accomplished around 30 min after the addition of the elicitor. Large (i. e., several m long) field-collected sporophytes of *L. hyperborea*, *L. ochroleuca* and *Saccorhiza polyschides* exhibited high constitutive rates of peroxide release, and were not responsive to the elicitor. The minimum size for *Laminaria digitata*, *Lessonia nigrescens* and *Macrocystis pyrifera* sporophytes to respond with an oxidative burst following incubation with alginate oligosaccharides was 1 mm. Smaller, embryonic sporophytes did not react. With the exception of *L. ochroleuca*, gametophyte cultures of Laminariales did not exhibit an oxidative response following incubation with alginate oligosaccharides.

Outside the Laminariales, the macroscopic thalli of only a few species were found to react by an oxidative burst upon challenge with alginate oligosaccharides. In the Fucales, all of the species tested constitutively released considerable amounts of hydrogen peroxide. Only *B. bifurcata*, *P. canaliculata*, *S. muticum* and, in some cases, *Fucus vesiculosus* exhibited an enhanced peroxide emission upon addition of oligoguluronates (Tab. 1, Fig. 1 B). In the Dictyotales and Sphacelariales, a slightly increased peroxide emission was also observed upon addition of oligoguluronates. Among the Desmarestiales, only *Desmarestia dudresnayi* reacted to incubation with the elicitor, yet with a marked oxidative response (Table 1).

Among the Ectocarpales with erect filamentous thalli, *Pylaiella littoralis* was unique in performing a relatively strong and rapid oxidative burst after addition of oligoguluronates (Fig. 1C). *Eurychasma*-infected cultures of *P. littoralis* maintained steady-state peroxide levels one order of magnitude higher than healthy cultures, and the oxidative burst response of infected cultures was several times stronger than that of the uninfected strain. The filamentous stages of *Chordaria flagelliformis*, *Elachista stellaris* and *Spermatochnus paradoxus* displayed weak rates of peroxide release, yet all three of them, and especially *S. paradoxus*, maintained fairly high peroxide levels in long-term cultures. They also exhibited a rapid, though comparatively weak, reactivity upon elicitor addition. No other species of Ectocarpales tested showed a burst, with no difference due to the presence or absence of virus infections. The Scytothamnales, Sporochnales, Syringodermatales, in which microthalli were examined, the erect filamentous thalli of Tilopteridales and Cutleriales as well as two filamentous species *incertae sedis*, *Microzonia velutina* and *Zosterocarpus oedogoniae*, did not exhibit an oxidative response to challenging with alginate elicitors eithers (Table 1).

Inhibition of the oxidative burst abolishes natural resistance of kelp sporophytes towards epiphytic bacteria

To further investigate the potential role of the oxidative response in the control of microorganisms by brown algae, sporophytes of Laminariales were subjected to loss-of-function experiments as follows. The capacity to perform an oxidative burst was suppressed with an NAD(P)H oxidase inhibitor, diphenylene iodonium (DPI), and thalli were then challenged by epiphytic bacteria. As shown in previous experiments with *L. digitata* (Küpper et al. 2001) and in Fig. 1A for *M. pyrifera*, this drug completely suppressed the oligoalginate-induced oxidative reaction in kelp sporophytes.

In a first series of experiments (Table 2), axenic sporophytes of *M. pyrifera* were inoculated with an epiphytic alginolytic bacterium previously isolated from *L. digitata*, *Pseudomonas alginovora*, strain XO 17 (Corre et al. 1989, Boyen et al. 1990a). Controls with bacteria-free *M. pyrifera* sporophytes confirmed that the treatment with DPI up to concentrations as high as 10 µM had no direct toxic effect on the plants. In the absence of DPI and upon co-cultivation with the bacterium, *M. pyrifera* sporophytes also survived for several weeks, without any signs of disease. In contrast, the DPI-treated axenic sporophytes were rapidly attacked by the bacterium, resulting in de-coloration within 3-5 days and, finally, complete disintegration at ca. 10 days after the addition of the drug and inoculation (Table 2).

Similar experiments were performed with non-axenic cultures of young sporophytes of *L. digitata*. Even after moderate DPI treatments (5 µM for 1 h), *L. digitata* plantlets were rapidly (within 3 days) attacked by their natural bacterial epiflora, displaying symptoms similar to the decaying of axenic, DPI-treated *Macrocystis* plantlets artificially inoculated with *Pseudomonas alginovora*. Similar results were also observed when *M. pyrifera* sporophytes were re-inoculated with their natural bacterial epiflora (not shown).

Development of a model pathosystem for the assessment of induced resistance in kelp sporophytes

In an attempt to identify a suitable pathogen for artificial infection experiments, kelp sporophytes were challenged with the oomycete *Eurychasma dicksonii*, which had been shown to infect members of many orders of brown algae, including both gametophytes and embryonic sporophytes of *L. digitata* (Müller *et al.* 1999). However, no artificial infection could be established with *E. dicksonii* in *L. digitata* sporophytes with a size greater than 1 mm. In contrast, two brown algal endophytes of the genus *Laminariocolax* (Ectocarpales), *Laminariocolax tomentosoides* and *L. macrocystis,* proved to produce infections in a reliable and reproducible manner. When *L. digitata* or *M. pyrifera* plantlets, from 2-20 mm in size, were inoculated with sporulating *Laminariacolax* filaments, the pathogens grew both endo- and epi-phytically (Figs 2A-B). *Laminariacolax* caused serious growth perturbations of host thalli, finally overgrowing its hosts, with dense tufts of *Laminariocolax* on the heavily infected plantlets (Fig. 2C), accompanied by massive necroses (Fig. 2D), and frequently resulting in thallus death. Under UV-light observation, the penetration or protrusion zones of infecting *Laminariocolax* filaments (Fig. 2E) showed an accumulation of bright-blue autofluorescent compounds (Fig. 2F). The first symptoms, consisting of local necroses, were visible within ten days of inoculation, but the extent of infection was best assessed at one month after inoculation.

Laminariocolax tomentosoides and *L. macrocystis* were found to be cross-infective, and attacked both *M. pyrifera* and *L. digitata*. Cell-free extracts of neither brown algal endophyte induced accumulation of H_2O_2 in *L. digitata* sporophyte cultures. Addition of the cell-free extracts 60 min prior to elicitation with 150 µg. mL^{-1} oligoguluronates instead considerably inhibited the oxidative burst (Fig. 3A). Cell-free extracts of *Laminariocolax* indeed exhibited a high peroxide-scavenging capacity. Typically, 100 µl extract per 10 mL of sea water degraded 10 µM hydrogen peroxide to over 95 % within less than 5 h, 50 % within the first 15 minutes. However, reactivity of *Laminaria* sporophytes was re-established after transfer to fresh medium lacking endophyte extract (Fig. 3B). In contrast, cell-free-extracts from the non-pathogenic ectocarpoid *Pylaiella littoralis* triggered an oxidative burst response in *L. digitata*. Noteworthily also, extracts of *Pylaiella littoralis* were twice as active when the alga was infected by the oomycete *Eurychasma dicksonii* (Fig. 1D).

Alginate oligosaccharides induce resistance in *Laminaria digitata* against *Laminariocolax tomentosoides*

The potential role of the oxidative burst in protecting *Laminaria* sporophytes from infections by brown algal endophytes was next addressed, by treating *L. digitata* plantlets with alginate oligosaccharides at the concentration of 150 µg. mL^{-1}, followed by challenge with *L. tomentosoides*. Non-elicited sporophytes inoculated with the endophyte displayed the symptoms described above (Figs 2, 4 A-D), i.e. dense infestation by endo- and epiphytic *Laminariocolax* filaments and conspicuous necrotic lesions. In contrast, young sporophytes which had been incubated for 2 hours in the presence of oligoguluronates, allowed to grow for 8 days in normal seawater medium and which were then challenged by *L. tomentosoides* sporulating filaments were remarkably well protected against infection. Very few – if any - epi- or endo-phytic filaments were visible and even fewer necroses occurred (Fig. 4 E-H). Remarkably also, whilst untreated plants showed a strong, localized accumulation of blue-autofluorescent compounds in the cortical cells around penetrating *Laminariocolax* filaments (Fig. 4 D), the oligoguluronate-treated plants exhibited a less intense, but ubiquitous cortical accumulation of blue-autofluorescent compounds (Fig. 4 H).

Using a relative scale to score the extent of infection and various time intervals between elicitation of *Laminaria digitata* thalli with alginate oligosaccharides and inoculation by *Laminariocolax tomentosoides*, we next investigated the time required for the induction of resistance and its duration. As shown in Figure 5, 8 days were required between elicitation and inoculation to observe a significant protection by alginate oligosaccharides, while the induced resistance was maintained for at least 6 more days. In other experiments, establishment of resistance could be observed as early as 3-5 days after elicitation (not shown).

Discussion

The oxidative burst is an essential component in natural or induced resistance in kelps

Several, convergent lines of evidence emerging from this study show that accumulation of reactive oxygen species, the so-called oxidative burst response, is an essential component of the defense system of kelp sporophytes against potentially harmful microorganisms. In loss-of-function experiments, in which the capacity of performing an oxidative burst was suppressed by the NADP(H) oxidase inhibitor diphenylene iodonium (Fig. 1 A), axenic sporophytes of *Macrocystis pyrifera* became very susceptible to the alginate-degrading, yet naturally non-pathogenic bacterium, *Pseudomonas alginovora* (Table 2). Similarly, control of the bacterial epiflora by non-axenic sporophytes of *Laminaria digitata* was shown to depend on a DPI-sensitive mechanism, because this kelp species is similarly capable of triggering an oxidative response following recognition of alginate oligosaccharides (Küpper et al. 2001). It follows that, in Laminariales, the capacity of releasing activated oxygen species appears as a central, causal element for keeping the bacterial epiflora at a tolerable level. Such a biochemical basis for the interaction between kelps and alginolytic bacteria is reminiscent of the ecological relationship between red algal agarophytes and their agar-degrading epiphytic bacteria, which also depends on the capacity of the host to perform an oxidative response following recognition of agar oligosaccharides (Weinberger and Friedlander 2000, Weinberger et al. 2001).

Reciprocally, in gain-of-function experiments by incubation with alginate oligosaccharides, elicited *L. digitata* sporophytes became resistant to infection by their natural endophytic pathogen *Laminariocolax tomentosoides*. A two-hour incubation in the presence of alginate oligosaccharides is known to induce a strong emission of H_2O_2 in *L. digitata* (Küpper et al. 2001), further indicating that the oligoguluronate-induced oxidative burst also plays a central role in the induction of resistance to the endophyte. This observation in kelp sporophytes is in agreement with the report that the oxidative burst in *Chondrus crispus* is a key element in establishing resistance to its green algal pathogenic endophyte *Acrochaete operculata* (Bouarab et al. 1999) and also with studies indicating a major role of H_2O_2 in the induction of systemic acquired resistance in higher plants (Alvarez et al. 1998, Van Camp et al. 1998).

Other observations in this study indirectly support the above conclusions. In natural populations, the sporophytes of Laminariales can show a near-100 % prevalence of pathogenic endophytes with severe disease symptoms such as warts, crippled thalli and twisted stipes (Ellertsdóttir & Peters 1997). We show here (Fig. 3) that, unlike extracts of the closely-related ectocarpoid, *Pylaiella littoralis*, *Laminariocolax* cell-free extracts not only are not recognized by *L. digitata* sporophytes but also are capable of quenching the alginate-induced oxidative response. This finding suggests that the great infective success of these endophytic pathogens may be due to their capability to avoid recognition or even to suppress the defenses of the host. In contrast to *Laminariocolax*, *Eurychasma dicksonii* is fended off by the sporophytes of Laminariales, whereas it is a natural pathogen to a large variety of brown algal groups (Küpper and Müller 1999, Müller et al. 1999). As suggested by Fig. 1D, this oomycete may produce compounds that are specifically recognized by *L. digitata*. Remarkably also, however, the virulence of this pathogen correlates well with the inability of its brown algal hosts to perform an oxidative burst. In particular, the threshold size (ca. 1 mm) above which infection by *Eurychasma* was no longer observed in *L. digitata* sporophytes is coincident with the threshold size for the acquisition of the capacity to respond with an oxidative burst to elicitation with alginate oligosaccharides.

Biological relevance and consequences of oligoguluronate recognition in Laminariales

In any host-pathogen interaction, recognition of the occurrence of an attack is the first line of defense. This may involve recognition of the attacker, through either the perception of exogenous elicitors, or of degradation products from the host itself (endogenous elicitors), or both. Therefore, in Laminariales, the oxidative response to the perception of alginate oligosaccharides is probably related to the sensing of damage to the alginate-rich cell wall matrix of kelps by microorganisms. Among the natural bacterial epiflora of Laminariales, many strains have been reported to harbor alginate-degrading enzymes. These include non-pathogenic bacteria, such as the one used in this study, *Pseudomonas alginovora* (Boyen et al. 1990a), and pathogenic strains, such as the agent of the spot-wounded disease in *L. japonica*, *Pseudomonas elyakovii* (Sawabe et al. 2000). No alginolytic activities have so far been reported in brown algal endophytes. However, scanning electron microscopy investigations of the early stages of the development of *Laminariocolax* and *Laminarionema* zoospores indicate that their mode of penetration may involve the enzymatic degradation of host surfaces rather than a mechanical disruption (Heesch and Peters 1999). Finally, alginases have been detected in marine fungi (Wainwright & Sherbrock-Cox 1981) as well as in a variety of kelp grazers, including molluscs and echinoderms (see Boyen et al. 1990b). On the other hand, oligoguluronate signalling may also be related to the need for recognizing challenge by alginate-containing brown algal endophytes and epiphytes.

As discussed above, the oxidative burst appears as one essential element of the biochemical basis of the immunity of Laminariales. This conclusion is consistent with previous findings showing that apoplastic peroxide concentrations such as those occurring during an oxidative burst can inhibit the growth of kelp-associated bacteria (Küpper et al. 2001). It suggests that activated oxygen species may have a direct role in controlling the growth of potentially pathogenic microorganisms. However, we show here that protection of *L. digitata* against *L. tomentosoides* was not fully established until at least 5 to 8 days after host elicitation with oligoguluronates and that it persisted for one more week (Fig. 5). These observations suggest that secondary, long-term defense mechanisms are induced downstream of the oxidative burst, a build-up of immunity which probably requires the progressive upregulation of biosynthetic enzymes as well as ultrastructural changes. We thus postulate that the establishment of resistance involves the synthesis of chemical defenses, which are either toxic to *Laminariocolax* zoospores and/or which result in the host cell wall strengthening, providing a barrier against penetration by the pathogen.

Chemical defenses of marine brown algae include secondary metabolites such as terpenoids and polyphenols (Hay and Fenical 1988 1992, Steinberg 1992, Hay and Steinberg 1992, Cronin and Hay 1996, Hay 1996, Pavia and Toth 2000), as well as fatty acid-derived compounds such as C_{11} hydrocarbons (Hay et al. 1998, Schnitzler et al. 1998). Interestingly, accumulation of UV-autofluorescent material was observed in the *L. digitata* plantlets elicited with alginate oligosaccharides as well as at the penetration sites of *Laminariocolax* filaments (Figs 4H and D, respectively), and we hypothesize that these UV-fluorescent compounds correspond to aromatic secondary metabolites such as phlorotannins. It was demonstrated that these defense metabolites are inducible by grazers in brown algae (Pavia and Toth 2000) and that this can increase resistance to further grazing (Toth and Pavia 2000). There is still some debate, however, as to the nature of the signals which induce the upregulation of phlorotannins. In Laminariales, mechanical wounding induces rapid accumulation of these compounds (Hammerstrom et al. 1998). In contrast, in the fucalean *Ascophyllum nodosum*, mechanical wounding did not properly mimic the action of the grazer *Littorina obtusata* regarding the build-up of phlorotannin defenses (Pavia and Toth 2000). In the light of our results, we propose that, besides mechanical wounding, signalling through alginate oligosaccharides may be a cue for the recognition of grazing by marine molluscs.

Phylogenic repartition of the elicitor-induced oxidative response in brown algae

Among the 11 brown algal orders under investigation here, high levels of constitutive release of hydrogen peroxide were observed in species which feature filamentous erect thalli or microthalli (Table 1). Such a behavior may constitute a non-enzymatic mechanism for the detoxification of activated oxygen species generated from photosynthesis or respiration (Collén and Davison 1998). The alginate-induced oxidative response was found in one member of the Ectocarpales and Desmarestiales each (*Pylaiella littoralis* and *Desmarestia dudresnayi*, respectively), and in the sporophytes of all of the Laminariales tested, whereas Fucoids, most Desmarestiales and filamentous species, including the gametophytes of Laminariales, were insensitive to challenging with oligoguluronates (at least respective to the oxidative response, Table 1).

Based on a recent phylogeny of brown algae (Rousseau et al. 2001), no clear evolutionary pattern can be drawn from the distribution of this trait in this survey. Noteworthily, however, it was found in one members of the Ectocarpales, in one of the Desmarestiales, and in the Laminariales, i. e., groups which have emerged the most recently as independent clades in the Phaeophyceae. With the exception of *P. littoralis*, the capacity of performing an alginate-induced oxidative burst may also be thought of as a developmentally-regulated characteristic of the alginate-rich members of brown algae with complex, foliose or frondose macrothalli, such as the sporophytes of Laminariales. Compared to the sporophytes, kelp gametophytes are very susceptible to infection by *Eurychasma dicksonii* (Müller *et al.* 1999) and the protist parasite *Maullinia ectocarpii* (Plasmodiophorales, Maier et al. 2000). The usually short life span and simple morphology of gametophytes, however, may have reduced the need for an inducible defense system. It can not be ruled out, however, that the macrothallic brown algae which do not respond to a challenge with oligoguluronates by a burst of H_2O_2 are probably endowed with other, inducible defense mechanisms. In this respect, the fucoid *Ascophyllum nodosum*, which was not responsive to oligoalginates (Table 1), was nevertheless recently shown capable of expressing inducible defenses against herbivores (Pavia and Toth 2000).

In conclusion, the capacity to recognize alginate oligosaccharides as defense signals is limited to a few members of brown algae. In the Rhodophytes, recognition as endogenous defense elicitors of host cell wall degradation products also appears as a rather specific response. In agar-containing red algae, recognition of agar oligosaccharides is so far restricted to the order Gracilariales (Weinberger et al. 1999), while in the red algal carrageenophyte

Chondrus crispus endogenous cell wall fragments are not recognized as defense signals (Bouarab et al. 1999). From a bioenergetic viewpoint, elicitor recognition followed by a massive, yet transient release of hydrogen peroxide appears as a rather economic (in terms of redox equivalents required for the reduction of oxygen) and finely tuned form of defense signalling.

In addition, our study shows that recognition of alginate oligosaccharides and the related oxidative response are involved with natural and induced immunity of kelps against epiphytic bacteria and endophytes. We also propose that alginate oligosaccharides can be thought of as a relevant defense cue in the context of herbivory. Regarding the ecology of marine plant communities, these views may be extended to the seaweeds capable of recognizing endogenous or exogenous cell wall fragments, such as the red algae mentioned above. Such a form of defense signalling is compatible with the conditions listed by Pavia and Toth (2000) for the emergence of non-constitutive defenses in plants. In any case, these elicitors can now be used to explore the biochemical basis of kelp immunity, including for identifying the chemical nature of the inducible defenses. In particular, the *Laminaria digitata-Laminariocolax tomentosoides* host-pathogen system employed in this study provides an experimental model to further investigate the regulation mechanisms and the biological activity of polyphenols in kelps.

Acknowledgements

We would like to thank Ingo Maier (University of Konstanz) for providing laboratory cultures of various algal species (Tab. 1). Claire Gachon (Ecole Normale Supérieure, Paris) and Doris Bliestle (Photographic laboratory, University of Konstanz) contributed to microscopy and photography of infected plants, respectively. F.C.K. was supported by fellowships from Studienstiftung des Deutschen Volkes (Bonn) and the European Commission (Program MAST-III).

References

Alvarez, M.E., R.I. Pennell, P.-J. Meijer, A. Ishikawa, R. Dixon, and C. Lamb. 1998. Reactive oxygen intermediates mediate a systemic network in the establishment of plant immunity. Cell **92**: 773-784.

Apostol, I., P.S. Low, P. Heinstein, R.D. Stipanovic, and D.W. Altman. 1987. Inhibition of elicitor-induced phytoalexin formation in cotton and soybean cells by citrate. Plant Physiology **84**: 1276-1280.

Apt, K. E.. 1988. Etiology and development of hyperplasia induced by *Streblonema* sp. (Phaeophyta) on members of the Laminariales (Phaeophyta). Journal of Phycology **24**: 28-34.

Bergey, D.R., M. Orozco-Cardenas, D.S. De Moura, and C.A. Ryan. 1999. A wound- and systemin-inducible polygalacturonase in tomato leaves. Proceedings of the National Academy of Sciences of the USA **96**: 1756-1760.

Boller, T. 1995. Chemoperception of microbial signals in plant cells. Annual Review of Plant Physiology & Plant Molecular Biology **46**: 189-214.

Bouarab, K., P. Potin, J. Correa, and B. Kloareg. 1999. Sulfated oligosaccharides mediate the interaction between a marine red alga and its green algal pathogenic endophyte. Plant Cell **11**: 1635-1650.

Boudart, G., C. Lafitte, J.C. Barthe, D. Frasez, and M.T. Esquerré-Tugayé. 1998. Differential elicitation of defense responses by pectic fragments in bean seedlings. Planta **206**: 86-94.

Boyen, C., Y. Bertheau, T. Barbeyron, and B. Kloareg. 1990a. Preparation of guluronate lyase from *Pseudomonas alginovora* for protoplast isolation in *Laminaria*. Enzyme and Microbial Technology **12**: 885-890.

Boyen, C., B. Kloareg B., M. Polne-Fuller, and A. Gibor. 1990b. Preparation of alginate lyases from marine molluscs for protoplast isolation in brown algae. Phycologia **29**(2): 173-181.

Brisson, L.F., R. Tenhaken, and C. Lamb. 1994. Function of oxidative cross-linking of cell wall structural proteins in plant disease resistance. Plant Cell **6**: 1703-1712.

Burkhardt, E., and A.F. Peters. 1998. Molecular evidence from nrDNA ITS sequences that *Laminariocolax* (Phaeophyceae, Ectocarpales *sensu lato*) is a worldwide clade of closely related kelp endophytes. Journal of Phycology **34**: 682-691.

Carpenter, R.C. 1990. Competition among marine macroalgae: a physiological perspective. Journal of Phycology **26**: 6-12.

Collén, J., and I.R. Davison. 1999. Reactive oxygen production and damage in intertidal *Fucus* spp. (Phaeophyceae). Journal of Phycology 35, 54-61

Correa, J. A. 1997. Infectious diseases of marine algae: Current knowledge and approaches. In: F. E. Round and D. J. Chapman (eds): Progress in Phycological Research. Vol. 12, Biopress Ltd. 149-180.

Corre, S., D. Prieur, S. Chamroux, J.-Y. Floc'h, and A. Hourmant. 1989. Caractérisation des communautés bactériennes épiphytes de frondes de *Laminaria digitata* et de débris résultant de leur fragmentation. Cahiers de Biologie Marine **30**: 115-130.

Côté, F., K.S. Ham, M.G. Hahn, and C.W. Bergmann. 1998. Oligosaccharides elicitor in host-pathogen interactions. Generation, perception and signal transduction. In: B.B. Biswas and H.K. Das (eds): Subcellular Biochemistry, Plant-Microbe Interactions, Vol. 29 (Plenum Press, New York), 385-432.

Cronin, G., and M.E. Hay. 1996. Induction of seaweed chemical defenses by amphipod grazing. Ecology **77**: 2287-2301.

Ebel, J., and A. Mithöfer. 1998. Early events in the elicitation of plant defence. Planta 206, 335-348

Ebel, J., A.A. Bhagwat, E.G. Cosio, M. Freger, U. Kissel, A. Mithöfer, and T. Waldmüller. 1995. Elicitor-binding proteins and signal transduction in the activation of a phytoalexin defense response. Canadian Journal of Botany **73** (suppl. 1): S506-S510.

Ellertsdóttir, E., and A.F. Peters. 1997. High prevalence of infection by endophytic brown algae in populations of *Laminaria* spp. (Phaeophyceae). Marine Ecology Progress Series **146**: 135-143.

Fritig, B., T. Heitz, and M. Legrand. 1998. Antimicrobial proteins in induced plant defense. Current Opinion in Immunology **10**: 16-22.

Fry, S.C., S. Aldington, P.R. Hetherington and J. Aitken. 1993. Oligosaccharides as signals and substrates in the plant cell wall. Plant Physiology **103**: 1-5.

Hammerstrom, K., M.N. Dethier, and D.O. Duggins. 1998: Rapid phlorotannin induction and relaxation in five Washington kelps. Marine Ecology Progress Series 165, 293-305

Hammond-Kossack, K.E. and J.D.G. Jones. 1996. Resistance gene-dependent plant defense responses. Plant Cell **8**: 1773-1791.

Haug, A., Larsen, B., and O. Smidsrød. 1974. Uronic acid sequence in alginate from different sources. Carbohydrate Research **32**: 217-225.

Hay, M.E. 1996. Marine chemical ecology: what's known and what's next? Journal of Experimental Marine Biology and Ecology **200**: 103-134.

Hay, M.E., and W. Fenical. 1988. Marine plant-herbivore interactions: The ecology of chemical defense. Annual Review of Ecology and Systematics **19**: 111-145.

Hay, M.E., and W. Fenical. 1992. Chemical mediation of seaweed-herbivore interactions. In "Plant-Animal Interactions in the marine benthos", ed. by D.M. John, S.J. Hawkins and J.H. Price, **46**: 319-337.

Hay, M.E., J. Piel, W. Boland, and I. Schnitzler. 1998. Seaweed sex pheromones and their degredation products frequently suppress amphipod feeding but rarely suppress sea urchin feeding. Chemoecology **8**: 91-98.

Hay, M.E., and P.D. Steinberg. 1992. The chemical ecology of plant-herbivore interactions in marine versus terrestrial communities. In: Rosenthal, G.A., Berenbaum, M. R. (ed.): Herbivores: their interactions with secondary plant metabolites. Academic Press, Inc., 371-413.

Heesch S., and A.F. Peters. 1999. Scanning electron microscopy of host entry by two brown algae endophytic in *Laminaria saccharina* (Laminariales, Phaeophyceae).- Phycological Research **47**: 1-5.

Kawai, H., and M. Tokuyama. 1995. *Laminarionema elsbetiae* gen. et sp. nov. (Ectocarpales, Phaeophyceae), a new endophyte in *Laminaria* sporophytes.- Phycological Research **43**: 185-190.

Koricheva, J., S. Larsson, and E. Haukioja. 1998. Insect performance on experimentally stressed woody plants: a meta analysis. - Annual Review of Entomology **43**: 195-216.

Küpper, F.C., and D.G. Müller. 1999. Massive occurrence of the heterokont and fungal parasites *Anisolpidium*, *Eurychasma* and *Chytridium* in *Pylaiella littoralis* (Ectocarpales, Phaeophyceae).- Nova Hedwigia **69**: 381-389.

Küpper, F.C., B. Kloareg, J. Guern, and P. Potin. 2001. Oligoguluronates elicit an oxidative burst in the brown algal kelp *Laminaria digitata*. Plant Physiology, **125**: 278-291.

Lüning, K., A. Wagner, and C. Buchholz. 2000. Evidence for inhibitors of sporangium formation in *Laminaria digitata* (Phaeophyceae) during the season of rapid growth. Journal of Phycology **36**: 1129-1134.

Maier, I., E. Parodi, R. Westermeier, and D.G. Müller. 2000. *Maullinia ectocarpii* gen. et sp. nov. (Plasmodiophorea), an intracellular parasite in *Ectocarpus siliculosus* (Ectocarpales, Phaeophyceae) and other filamentous brown algae. Protist **151**: 225-238.

Müller, D.G., F.C. Küpper, and H. Küpper. 1999. Infection experiments reveal broad host ranges of *Eurychasma dicksonii* (Oomycota) and *Chytridium polysiphoniae* (Chytridiomycota), two eukaryotic parasites in marine brown algae (Phaeophyceae).- Phycological Research **47**: 217-23.

Northup, R.R., R.A. Dahlgren and J.G. McColl. 1998. Polyphenols as regulators of plant-litter-soil interactions in northern California's pygmy forest: A positive feedback? Biogeochemistry **42**: 189-220

Norton, T.A., D.C. Watson, S.J. Hawkins, N.L. Manley, and G.A. Williams. 1990. Scraping a living: a review of littorinid grazing. Hydrobiologia **193**: 117-38

Nürnberger, T., D. Nennstiel, T. Jabs, W.R. Sacks, K. Hahlbrock, and D. Scheel. 1994. High affinity binding of a fungal oligopeptide elicitor to parsley plasma membranes triggers multiple defense responses. Cell **78**: 449-460.

Olff, H., and M.E. Ritchie. 1998. Effects of herbivores on grassland plant diversity. Trends in Ecology and Evolution **13:** 261-265

Otte, O., and W. Barz. 1996. The elicitor-induced oxidative burst in cultured chickpea cells drives the rapid insolubilization of two cell wall structural proteins. Planta **200**: 238-246.

Pavia, H., and G. Toth. 2000. Inducible chemical resistance in the brown seaweed *Ascophyllum nodosum*. Ecology **81**: 3212-3225.

Pellegrini, M., and L. Pellegrini. 1982. Some observations on relationships between bacteria and a brown alga. Biology of the Cell **43**: 195-200.

Peñuelas, J. and M. Estiarte 1998. Can elevated CO_2 affect secondary metabolism and ecosystem function? Trends in Ecology and Evolution 13: 20-24.

Peters, A.F. 1991. Field and culture studies of *Streblonema macrocystis* sp. nov. (Ectocarpales, Phaeophyceae) from Chile, a sexual endophyte of giant kelp. Phycologia **30**(4): 365-377.

Peters, A.F. and E. Ellertsdóttir. 1996. New record of the kelp endophyte *Laminarionema elsbetiae* (Phaeophyceae, Ectocarpales) at Helgoland and its life history in culture. Nova Hedwigia **62**: 341-349.

Peters, A.F. and M.E. Ramirez 2001. Molecular phylogeny of small brown algae, with special reference to the systematic position of *Caepidium antarcticum* (Adenocystaceae, Ectocarpales). Cryptogamie, Algologie, **22**: in press

Peters, A.F. and B. Schaffelke. 1996. *Streblonema* (Ectocarpales, Phaeophyceae) infection in the kelp *Laminaria saccharina* (Laminariales, Phaeophyceae) in the western Baltic. Hydrobiologia **326/327**: 111-117.

Potin, P., K. Bouarab, F.C. Küpper, and B. Kloareg. 1999. Oligosaccharide recognition signals and defence reactions in marine plant-microbe interactions. Current Opinion in Microbiology **2**: 276-83.

Ricci, P., P. Bonnet, P., J.-C. Huet, M. Sallantin, F. Beauvais-Cante, M. Bruneteau, V. Billard, G. Michel, and J.-C. Pernollet. 1989. Structure and activity of proteins from pathogenic fungi *Phytophthora* eliciting necrosis and acquired resistance in tobacco.- European Journal of Biochemistry **183**: 555-563.

Rousseau, F., R. Burrowes, A.F. Peters, R. Kuhlenkamp, and B. de Reviers. 2001. A comprehensive phylogeny of the Phaeophyceae based on nrDNA sequences resolves the earliest divergences. Comptes Rendus de l'Académie des Sciences **324**: 1-15

Sawabe, T., R. Tanaka, M.M. Iqbal, K. Tajima, Y. Ezura, E.P. Ivanova, and R. Christen. 2000. Assignment of *Alteromonas elyakovii* KMM 162^T and five strains from spot-wounded fronds of *Laminaria japonica* to *Pseudoalteromonas elyakovii* comb. nov. and the extended description of the species.- International Journal of Systematic and Evolutionary Microbiology **50**: 265-71.

Scheel, D., and J.E. Parker. 1990. Elicitor recognition and signal transduction in plant defense gene activation. Zeitschrift für Naturforschung **45**: 569-575.

Scheel, D. 1998. Resistance response physiology and signal tranduction. Current Opinion in Plant Biology **1**: 305-310.

Schnitzler, I., W. Boland, and M.E. Hay. 1998. Organic sulfur compounds from *Dictyopteris* spp. deter feeding by an herbivorous amphipod (*Ampithoe longimana*) but not by an herbivorous sea urchin (*Arbacia punctulata*). Journal of Chemical Ecology **24**: 1715-1732.

Somssich, I.E., and K. Halbrock. 1998. Pathogen defense in plants: a paradigm of biological complexity. Trends in Plant Science **3**: 86-90.

Starr, R.C., and J.A. Zeikus. 1987. UTEX - the culture collection of algae at the University of Texas at Austin. Journal of Phycology **23** (Suppl.): 1-47

Steinberg, P. D. 1992. Geographical variations in the interaction between marine herbivores and brown algal secondary metabolites. In: V.J. Paul (ed.): Ecological roles of marine natural products. Cornell University, New York, USA, 51-92.

Uppalapati, S.R., and Y. Fujita. 2000. Carbohydrate regulation of attachment, encystment, and appressorium formation by *Pythium porphyrae* (Oomycota) zoospores on *Porphyra yezoensis* (Rhodophyta). Journal of Phycology **36**: 359-66

Toth, G.B., and H. Pavia. 2000. Water-borne cues induce chemical defense in a marine alga (*Ascophyllum nodosum*). Proceedings of the National Academy of Sciences of the USA **97**: 14418-14420.

Van Camp, W.V., M. Van Montagu, and D. Inzé. 1998. H_2O_2 and NO: redox signals in disease resistance. Trends in Plant Science **3**: 330-334.

Viard, M.-P., F. Martin, A. Pugin, P. Ricci, and J.-P. Blein. 1994. Protein phosphorylation is induced in tobacco cells by the elicitor cryptogein. Plant Physiology **104**: 1245-1249.

Wainwright, M., and U. Sherbrock-Cox. 1981. Factors influencing alginate degradation by the marine fungi: *Dendryphiella salina* and *D. arenaria*. Botanica Marina **24**: 489-91

Weinberger, F., M. Friedlander, and H.-G. Hoppe. 1999. Oligoagars elicit an oxidative burst in *Gracilaria conferta* (Rhodophyta). Journal of Phycology **35**: 747-755.

Weinberger, F., and M. Friedlander. 2000. Response of *Gracilaria conferta* (Rhodophyta) to oligoagars results in defense against agar-degrading epiphytes.- Journal of Phycology **36**: 1079-1086.

Weinberger, F., C. Richard, B. Kloareg, Y. Kashman, H.G. Hoppe, and M. Friedlander. 2001. Structure-activity relationships of oligoagar elicitors toward *Gracilaria conferta* (Rhodophyta). Journal of Phycology **37**: 1-9.

Yumoto, I., Y. Ezura, and T. Kimura. 1989. Distribution of *Alteromonas* sp., the causative agent of red-spots on the culture bed of Makonbu *Laminaria japonica*, in the coastal area of Funka Bay.- Nippon Suisan Gakkaishi **55**(3): 453-462.

ZoBell, C.E., and H.C. Upham. 1944. A list of marine bacteria including descriptions of sixty new species. Bulletin of the Scripps Institution of Oceanography of the University of California (Technical Series) **5**: 239-292.

Table 1. Survey of brown algal species, regarding their rates of release of hydrogen peroxide in the absence or the presence of oligoguluronates (150 µg. mL^{-1}). Basal hydrogen peroxide release was determined by following the concentration in dishes of control plants over 3 h, whilst permanent concentrations refer to measurements in the culture vessels in which the algae had been kept for at least one week. In some cases, culture medium from a stock culture collection was examined, in which algal samples were untouched for a minimum of one year in 4 ml of culture medium. In cases where no oxidative response to alginate oligosaccharides was observed for all the tested species for a given genus, at least one species was also elicited with valinomycin, which had been shown to activate the burst-producing machinery downstream of GG recognition (Küpper *et al.*, 2001). The classification of Ectocarpales *sensu lato* is based on Peters and Ramirez (2001). Abbreviations: gam = gametophytes; n.d. = not determined; sp = sporophytes. Origin of plant material / culture collections: [1)] Akira F. Peters, Kiel; [2)] Dieter G. Müller, Konstanz; [3)] Ingo Maier, Konstanz

Algal species	**Life history stage, morphology**	**Permanent H_2O_2 concentration in cultures**		**Basal H_2O_2 release**	**Occurrence of the oxidative burst**	**Origin of plant material**
		Culture dishes	**Medium of stock collection**			
Laminariales						
Chorda filum	macrothalli, sp.	n.d.		-	++	Galicia[1]
"	microthalli, gam	♂: 0.72 ± 0.12 µM ♀: 1.4 ± 1.2 µM	n.d.	-	-	Helgoland[3]
Laminaria digitata	sp	0.048 ± 0.26 µM		-	++	Roscoff (field-collected old sporophytes); Helgoland[2]

Table 1 continued.

Laminaria digitata	gam	♂: 0.12 ± 0.0008 µM ♀: 0.07 ± 0.002 µM	♂: 0.019 ± 0.0003 µM ♀: 0.004 ± 0.001 µM	-	-	Helgoland[2]
Laminaria hyperborea	sp	n.d.		field-coll.: ++ culture: -	++	Roscoff (field-collected old sporophytes; young sporophytes cultured from spore release of sori of field-collected plants)
Laminaria ochroleuca	sp	0.11 ± 0.24 µM		field-coll.: ++ culture: +-	++	Roscoff (field-collected old sporophytes; young sporophytes cultured from spore release of sori of field-collected plants)
"	gam	♀: 0.26 ± 0.22 µM ♂: 0.38 ± 0.25 µM	n.d.	-	+-	Roscoff[3]
Laminaria pallida	sp	n.d.	n.d.	-	++	Cape Peninsula, South Africa (cell suspension culture produced from a callus)[3]
	gam	n.d.	n.d.	-	-	Cape Peninsula, South Africa [3]
Laminaria saccharina	sp	n.d.		-	++	Roscoff (old sporophytes, field-collected), Helgoland (young sporophytes)
Macrocystis pyrifera	sp	n.d.		-	++	Fuerte Bulnes, Chile[2]
Lessonia nigrescens	sp	n.d.		-	++	Chiloe, Chile[2]
Saccorhiza polyschides	sp			++ (field-collected)	+	Roscoff (field-collected)
Fucales						
Ascophyllum nodosum	thallus fragments			++	+-	Roscoff (field-collected)

Table 1 continued.

Bifurcaria bifurcata	thallus fragments			++	+-	Roscoff (field-collected)
Cystoseira baccata	thallus fragments			++	-	Roscoff (field-collected)
Fucus evanescens	thallus fragments			++	-	Kiel (field-collected)
Fucus serratus	thallus fragments			++	-	Roscoff (field-collected)
Fucus vesiculosus	thallus fragments			++	+-	Kiel (field-collected)
Halidrys siliquosa	thallus fragments			++	-	Roscoff (field-collected)
Himanthalia elongata	thallus fragments			++	-	Roscoff (field-collected)
Pelvetia canaliculata	thallus fragments			+	+	Roscoff (field-collected)
Sargassum muticum	thallus fragments			++	+-	Roscoff (field-collected)
Desmarestiales						
Desmarestia aculeata	sp	0.64 ± 0.04 µM		++	+-	Helgoland[1,2]
Desmarestia dudresnayi	sp	n.d.		-	++	Galicia[1]
Desmarestia ligulata	sp	n.d.		++	-	Galicia[1], Roscoff (field-collected)
Desmarestia munda	gam	0.932 ± 0.1 µM	0.09 ± 0.037 µM	++	-	Bamfield, B.C., Canada[2]

Table 1 continued.

Scytothamnales						
Splachnidium rugosum	microthalli	4.6 ± 0.01 µM	0.015 ± 0.001 µM	+	-	Tasmania[2]
Dictyotales						
Dictyota dichotoma	sterile			+	+-	Roscoff (field-collected)
Dictyota phlyctaenodes	sterile	0.57 ± 0.17 µM	1.28 ± 0.04 µM	+	-	Juan Fernandez Islands[2]
Cutleriales						
Cutleria multifida	gam (♀)	1.33 ± 0.079 µM	0.9 ± 0.025 µM	+	-	Villefranche[2]
Sporochnales						
Sporochnus pedunculatus	gam	0.98 ± 0.04 µM	0.18 ± 0.02 µM	+	-	Villefranche, France, Mediterranean Sea[2]
Tilopteridales						
Tilopteris mertensii	non-sexual	0.9 ± 0.02 µM	0.14 ± 0.034 µM	+	-	Helgoland[2]
Syringodermatales						
Syringoderma phinneyi	gam	0.35 ± 0.03 µM	0.006 ± 0.0002 µM	+-	-	British Columbia, Canada[2]
Sphacelariales						
Cladostephus spongiosus	n.d. / sterile			+	-	Roscoff (field-collected)

Table 1 continued.

Sphacelaria rigidula	sp	0.27 ± 0.016 μM	0.007 ± 0.001 μM	+	-	California[2]
(*incertae sedis*)						
Microzonia velutina	microthalli	0.69 ± 0.003 μM	0.2 ± 0.01 μM	++	-	Portobello, New Zealand[2]
Zosterocarpus oedogoniae	unknown	0.81 ± 0.014 μM	0.06 ± 0.0001 μM	++	-	Villefranche, France, Mediterranean Sea[2]
Ectocarpales sensu lato						
Adenocystis utricularis	macrothalli	0.9 ± 0.03 μM	n.d.	++	-	Ushuaia, Argentina[1]
Chordaria flagelliformis	microthalli	0.074 ± 0.009 μM	0.077 ± 0.024 μM	-	+-	Helgoland[1]
Ectocarpus fasciculatus	partheno-sp. (♀), filamentous	0.17 ± 0.006 μM	0.14 ± 0.07 μM	+	-	Roscoff[2]
"	sp, filamentous	0.36 ± 0.01.5 μM	0.0035 ± 0.0002 μM	+	-	Ireland[2]
" + virus	sp, virus-infected, sterile, filamentous	2.39 ± 0.1.35 μM	0.002 ± 0.0004 μM	+	-	Roscoff[2]
" + virus	sp, sterile, virus-infected, filamentous	0.4 ± 0.04 μM	0.0018 ± 0.0001 μM	+	-	Ireland[2]
Ectocarpus siliculosus	gam (♂), filamentous	1.25 ± 0.116 μM	0.318 ± 0.056 μM	+	-	Kaikura, New Zealand[2]

Table 1 continued.

Ectocarpus siliculosus + virus	gam, filamentous	5.2 ± 0.069 µM	0.61 ± 0.197 µM	+	-	Kaikura, New Zealand[2]
"	sp, filamentous	0.0049 ± 0.005 µM	0.045 ± 0.004 µM	+	-	Port Aransas, Texas[2]
Elachista stellaris	macrothalli	0.3 ± 0.1 µM	0.13 ± 0.005 µM	-	+-	Villefranche, France, Mediterranean Sea[2]
Eudesme virescens	microthalli	2.8 ± 0.25 µM	0.18 ± 0.05 µM	+-	-	Aran Islands, Ireland[2]
Feldmannia irregularis	non-sexual, filamentous	1.03 ± 0.14 µM	0.712 ± 0.016 µM	+	-	Canary Islands[2]
" + virus	non-sexual, virus-infected, sterile, filamentous	1.46 ± 0.28 µM	0.866 ± 0.017 µM	+	-	Canary Islands[2]
Feldmannia simplex	non-sexual, filamentous	1.44 ± 0.0008 µM	0.022 ± 0.0072 µM	+	-	Aran Islands, Ireland[2]
" + virus	non-sexual, filamentous	0.8 ± 0.008 µM	0.013 ± 0.001 µM	+	-	Aran Islands, Ireland[2]
Myriotrichia clavaeformis	macrothalli	3.98 ± 0.06 µM	0.16 ± 0.074 µM	-	-	Las Grutas, Argentina[2]
" + virus	macrothalli, sterile, virus-inf.	1.9 ± 0.02 µM	0.046 ± 0.01 µM	-	-	Las Grutas, Argentina[2]
Petalonia fascia	microthalli	1.7 ± 0.2 µM	0.014 ± 0.005 µM	+-	-	Villefranche, France, Mediterranean Sea[2]
Pylaiella littoralis	non-sexual, filamentous	0.11 ± 0.02 µM	0.104 ± 0.06 µM	-+	+	Isla Diego Ramirez, Chile[2]
Pylaiella littoralis + *Eurychasma dicksonii*	non-sexual, filamentous	1.66 ± 0.009 µM		-	++	Shetland (pathogen), Isla Diego Ramirez (host)[2]

Table 1 continued.

Pylaiella littoralis + virus	sterile, virus-infected, filamentous	0.53 ± 0.005 µM	0.07 ± 0.003 M	+	+-	Isla Diego Ramirez (virus-infected in lab culture)[2]
Scytosiphon lomentaria	microthalli, gam (♀)	0.6 ± 0.08 µM	0.28 ± 0.04 µM	+	-	Victoria, Australia[2]
Spermatochnus paradoxus	microthalli	3.33 ± 0.11 µM	n.d.	-	+-	Villefranche, France, Mediterranean Sea[2]

Table 2. Effect of the NAD(P)H inhibitor diphenylene iodonium (DPI) on the interaction between *Macrocystis pyrifera* and the epiphytic bacterium *Pseudomonas alginovora* / XO 17. The observations recorded here are representative for 7 experiments with similar results.

+ and - refer to the presence or absence of inoculum, respectively. n: number of plants tested for a given condition.

Treatment	Bacterial challenge	n	Observations
axenic, untreated control	-	20	normal pigmentation
	+	20	normal pigmentation
1 µM DPI, 10 min	-	5	plants normal, few apically bleached
	+	5	plants seriously damaged, except a few stipes & meristems completely bleached
2.5 µM DPI, 10 min	-	5	normal pigmentation and growth
	+	5	plants entirely bleached except a few stipes and meristems
5 µM DPI, 10 min	-	5	some plants are bleached (except stipes), one half in normal condition
	+	5	all plants bleached except stipes, some are less damaged than those exposed to 10 µM DPI + bacteria
10 µM DPI, 10 min	-	5	plants in good shape, normal pigmentation
	+	5	strongly bleached, incl. stipes

Figures

Figure 1. Monitoring of hydrogen peroxide concentrations in the culture medium of various brown algae, in the presence or in the absence of alginate oligosaccharides (GG blocks, 150 µg. mL^{-1}). Addition of elicitors is marked by an arrow. A) *Macrocystis pyrifera* sporophytes (Laminariales) : GG only, diamonds; 1 µM DPI, squares; 2.5 µM DPI, triangles; 5 µM DPI, asterisks. Note that DPI suppressed the GG-triggered oxidative burst; B) *Fucus vesiculosus* (Fucales) : control, diamonds; GG, squares; C) *Pylaiella littoralis* (Ectocarpales), control, diamonds; GG, squares. D) *Laminaria digitata* (Laminariales), challenged with 150 µg. mL^{-1} GG (hollow circles), cell-free extracts from healthy *Pylaiella littoralis* (triangles) or from *Pylaiella littoralis* infected with *Eurychasma dicksonii* (squares).

Figure 2. Infection of *Laminaria digitata* by the endophyte *Laminariocolax tomentosoides* (all scale bars correspond to 100 µm). Superficial views of *Laminariocolax*-infected *L. digitata* plants showing that the pathogen lives both epiphytically (A) and endophytically, as filaments in the intercellular spaces (A, B). C) On heavily infected plantlets, dense tufts of *Laminariocolax* are visible. D) under UV excitation massive necroses are visible, as regions with bright-blue autofluorescence. E, F) Section of a strongly infected *L. digitata* plantlet, densely covered with *Laminariocolax* filaments. Penetration of *Laminariocolax* filaments is often accompanied by accumulation of bright-blue autofluorescent compounds in the surrounding cortical host cells. *Laminariocolax* filaments in the medulla of this *Laminaria* plant are visible by their darker red plastid fluorescence compared to the brighter red fluorescence of *Laminaria*.

Fig. 3. *Laminariocolax tomentosoides* cell-free extracts scavenge the reactive oxygen species generated by the oxidative burst in *Laminaria digitata* (addition of 150 µg. mL^{-1} GG is marked by an arrow). A) *Laminariocolax* extracts added 1 min after GG elicitation, asterisks; control, squares; GG only, triangles. B) Cell-free extracts were added for 45 min, medium was changed completely, and the elicitors were added 15 min later. In contrast to above, the treated plants (asterisks) were unhindered in their oxidative burst response.

Figure 4. Oligoguluronates induce resistance to *L. tomentosoides* in *L. digitata*. In controls (A-D), plantlets are completely overgrown (A, B), and frequently penetrated by (C, D) by *Laminariocolax* filaments. As shown also in Fig. 2, penetrating filaments are often surrounded by bright-blue autofluorescent compounds (D). *Laminaria digitata* plantlets were elicited with alginate oligosaccharides for 2 h with 150 µg. mL^{-1} GG, grown for 9 days in PES medium, then inoculated with *Laminariocolax tomentosoides* (E-H). Hardly any epi- or endophytic *Laminariocolax* filaments were visible, and the cortex of GG-treated plants often showed accumulation of autofluorescent compounds, yet less intense and much more widespread and diffuse than in untreated plants. Photographs were taken 34 days after inoculation with the pathogen. Scale bars: A – 1 mm; B, F: 1 cm; C, D, E, G, H: 10 µm

Figure 5. Time scale of the establishment and maintenance of the oligoguluronate-induced resistance against *Laminariocolax tomentosoides* in *Laminaria digitata.* Plants were elicited, then inoculated at various intervals, as shown in the x-axis. At 3 weeks after inoculation, relative marks of resistance, from 0 (maximum infestation) to 6 (complete resistance, see the Materials and Methods section), were given by an independent observer. Error bars indicate variability of resistance, expressed as the standard deviation of the marks attributed to a given series of plants.

Figure 1

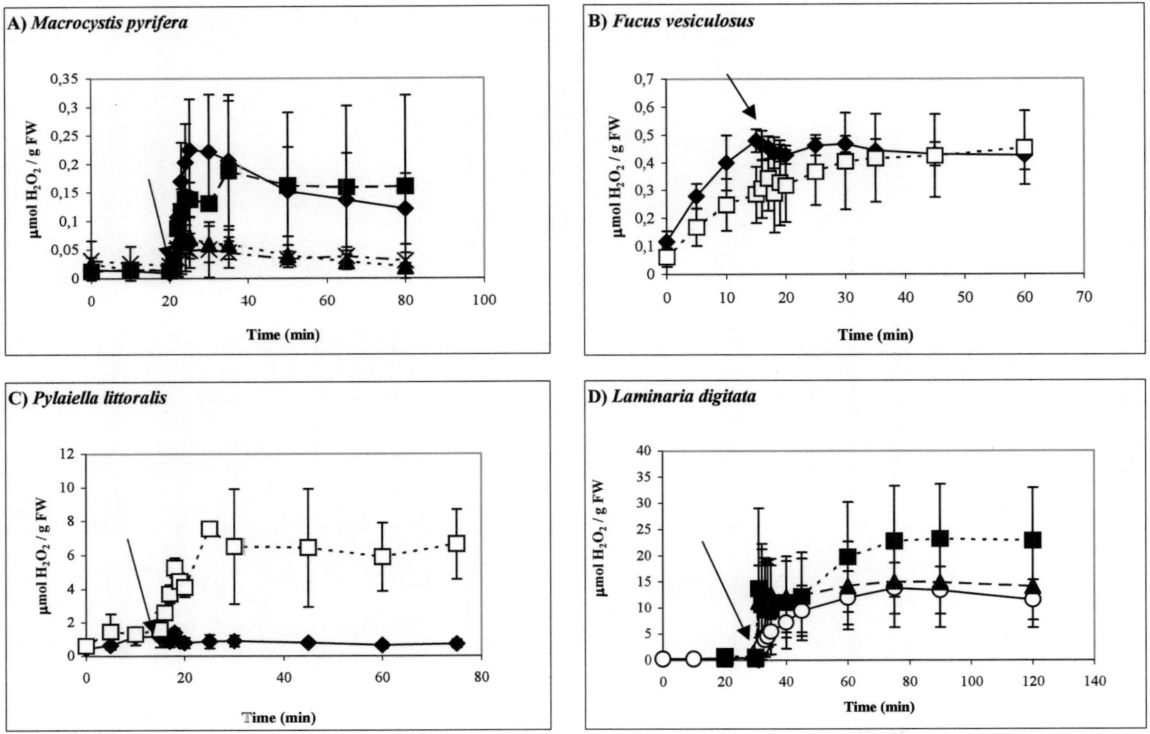
A) Macrocystis pyrifera
0,35
0,3
0,25
0,2
0,15
0,1
0,05
0
µmol H2O2 / g FW
0
20
40
60
80
100
Time (min)
B) Fucus vesiculosus
0,7
0,6
0,5
0,4
0,3
0,2
0,1
0
µmol H2O2 / g FW
0
10
20
30
40
50
60
70
Time (min)
C) Pylaiella littoralis
12
10
8
6
4
2
0
µmol H2O2 / g FW
0
20
40
60
80
Time (min)
D) Laminaria digitata
40
35
30
25
20
15
10
5
0
µmol H2O2 / g FW
0
20
40
60
80
100
120
140
Time (min)

Figure 2

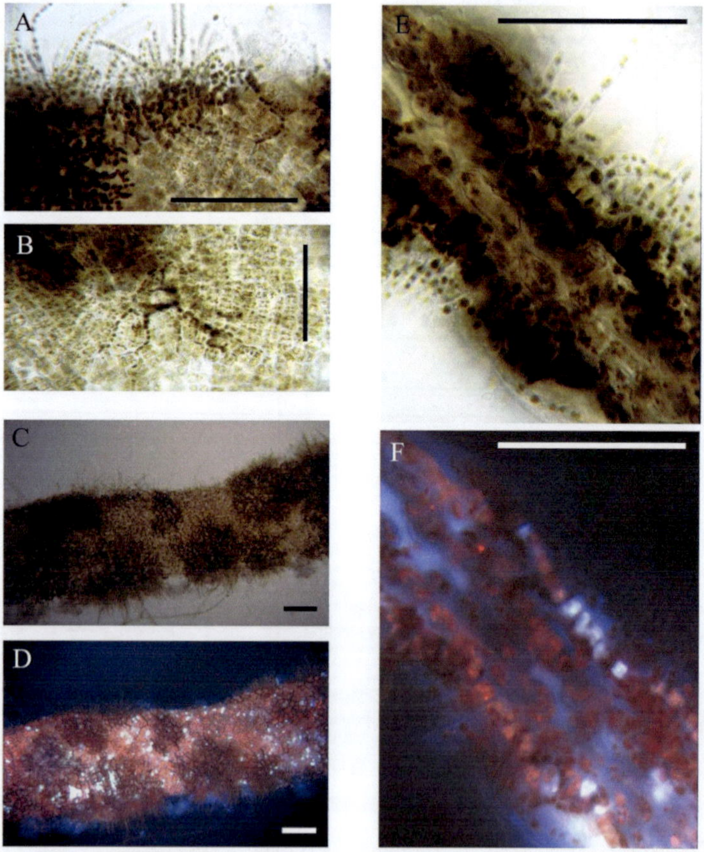

Figure 3

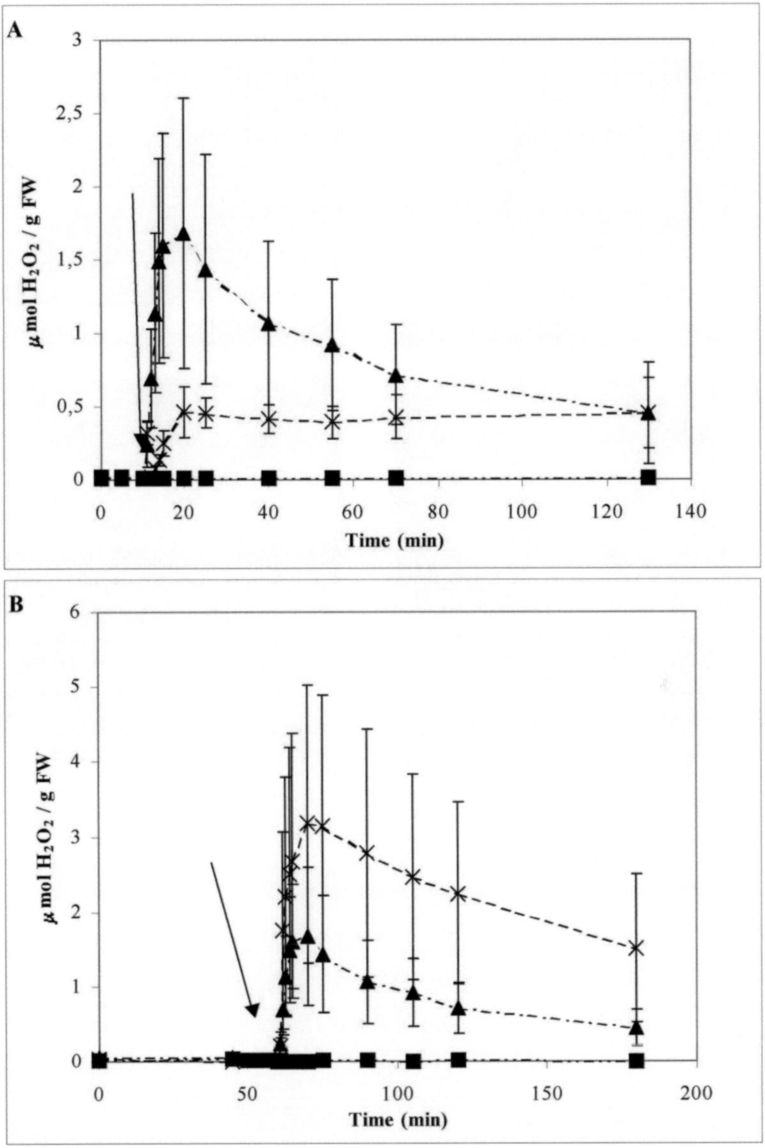
A
3
2,5
2
1,5
1
0,5
0
0
20
40
60
80
100
120
140
Time (min)
µ mol H_2O_2 / g FW
B
6
5
4
3
2
1
0
0
50
100
150
200
Time (min)
µ mol H_2O_2 / g FW

Figure 4

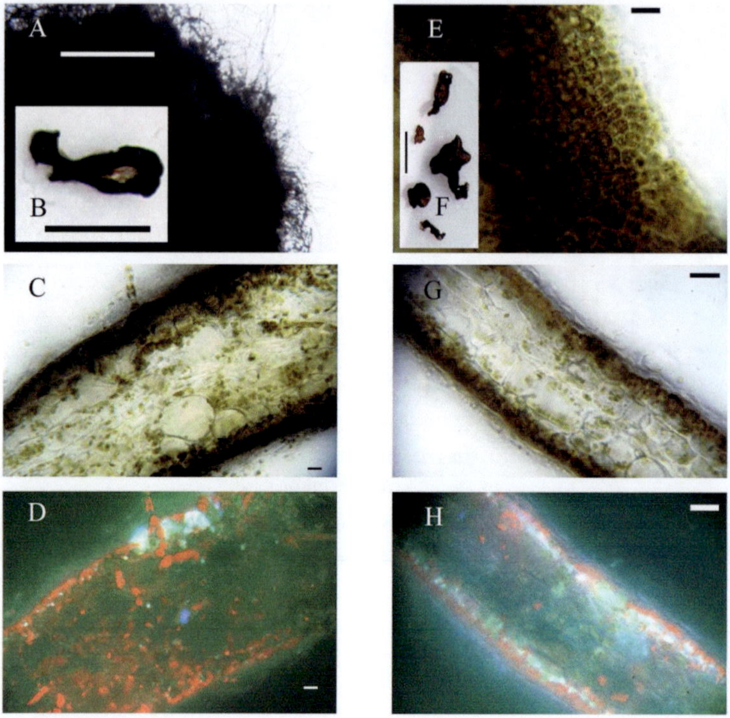

Figure 5

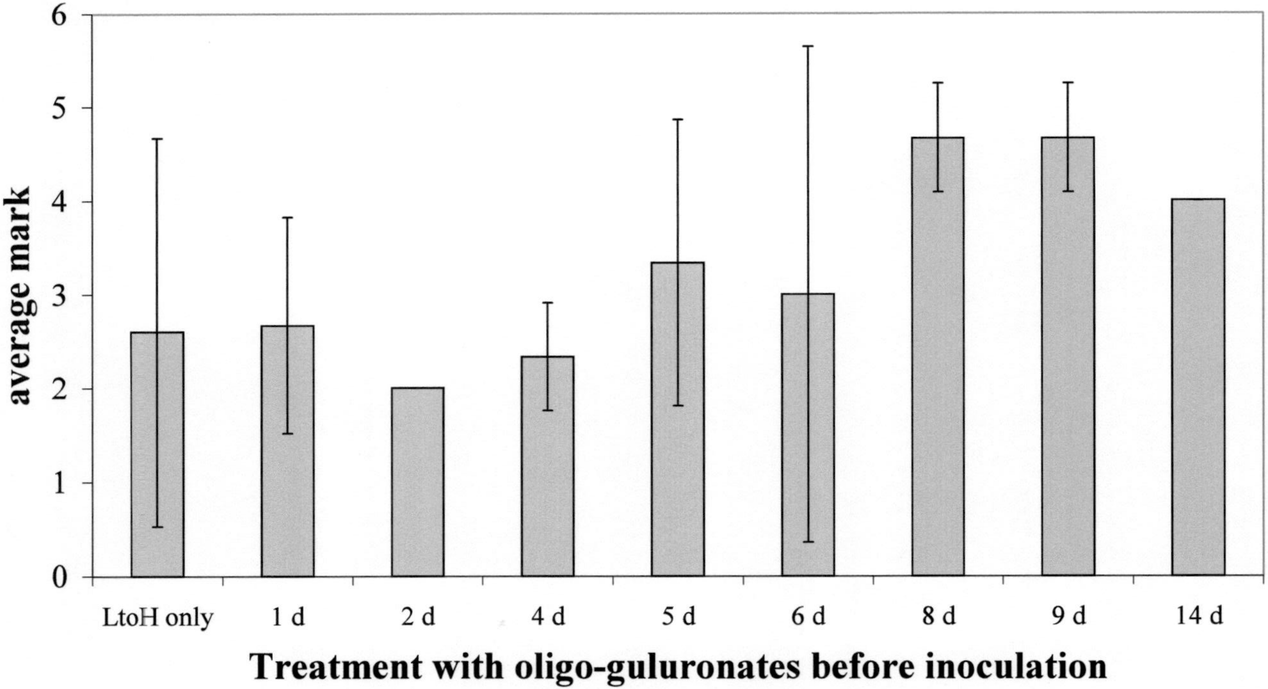

average mark
6
5
4
3
2
1
0
LtoH only
1 d
2 d
4 d
5 d
6 d
8 d
9 d
14 d
Treatment with oligo-guluronates before inoculation

10

Oxylipins trigger an oxidative burst and induce resistance in *Laminaria digitata* against the pathogenic endophyte, *Laminariocolax tomentosoides*

Frithjof Küpper[1,2], Fadi Adas[1, 3], Dieter G. Müller[2], Philippe Potin[1], Bernard Kloareg[1], Jean-Pierre Salaun[1, 3]

[1)]Station Biologique, UMR 1931 CNRS-Laboratoires Goëmar, BP 74, F-29682 Roscoff, Brittany, France

[2)]University of Konstanz, Dept. of Biology, D-78457 Konstanz, Germany

[3)]Laboratoire de Biochimie - EA-948, Université de Bretagne Occidentale, Faculté de Médecine, 22 Avenue C. Desmoulins, BP-815, F-29285 Brest, France

Abstract

This study provides first evidence that lipid oxidation signalling plays a key role in defense induction in marine brown algae. Arachidonic and linolenic acid, methyl jasmonate, prostaglandin A_2, various lipopolysaccharides and lipoteichoic acids from bacterial pathogens were found to be strong triggers of an oxidative burst in the kelp *Laminaria digitata*. Two different sources appear to be involved in the generation of reactive oxygen species: The first one is activated by oligosaccharides, arachidonic acid, methyl jasmonate and bacterial elicitors and can be inhibited by diphenylene iodonium (DPI). In contrast, the other pathway was found to be activated only by prostaglandin A_2, and could not be inhibited by DPI.

Furthermore, treatment of kelp plants with arachidonic acid and methyl jasmonate induced resistance against a pathogenic endophyte, *Laminariocolax tomentosoides*.

Abbreviations

DPI = diphenylene iodonium; FW = fresh weight; GG = oligo-guluronates; MeJA = methyl jasmonate; LPS = Lipopolysaccharide; LTA = Lipoteichoic acid; PGA_2 = prostaglandin A_2; ROS = reactive oxygen species

Introduction

In their responses to biotic and abiotic stress, cells are under the control of a plethora of regulatory molecules, which govern their actions to provide protection against toxicity and infection. Polyunsaturated fatty acids and their oxidized derivatives play a role in the signalling and production of hormones involved in both plant and animal stress responses (Farmer, 1994; Farmer *et al.*, 1998). Similar to mammals, biotic and abiotic plant stresses such as infection and exposure to xenobiotics induce an oxidation cascade of fatty acids leading to the synthesis of stress hormones (Blechert *et al.*, 1995).

Most unsaturated fatty acids in higher plants belong to the C_{18} family, whilst polyunsaturated C_{20} fatty acids seem to occur only in very few species. The mammalian response to microbial attack involves the activation of immune response genes, including those involved in the arachidonic acid cascade. In comparison, plants synthesize compounds of direct antimicrobial toxicity (reactive oxygen species / ROS, phytoalexins, pathogenesis-related proteins) and activate genes involved in the oxylipin cascade such as lipoxygenases, in a process that appears similar to the arachidonic acid cascade in mammals. Many inducible defense genes are regulated by signalling pathways involving one or more low molecular weight signal molecules, such as jasmonic acid, salicylic acid, ethene, and possibly H_2O_2 (Reymond & Farmer, 1998; Orozco-Cárdenas *et al.*, 2001).

The oxidative burst which appears both as a rapid defense response and an internal emergency signal, consists of the rapid and massive production of reactive oxygen species (ROS) such as superoxide (O_2^-), hydrogen peroxide (H_2O_2), and hydroxyl radicals ($^{\cdot}OH$) which are likely produced via the interaction of plasma membrane-associated NAD(P)H oxidases (Wojtaszek, 1997). Hence, pathogen attack triggers both the production of ROS and the oxidation cascade of fatty acids leading to bioactive oxylipins in plants.

Similarly, the understanding of conserved principles of pathogen recognition in animals, terrestrial plants and other groups of eukaryotes is beginning to emerge. In recent years, a body of evidence was gathered supporting the theory of so-called pattern-recognition receptors (Pugin *et al.*, 1994), including Toll-like receptors (Lien *et al.*, 1999) for a wide range of bacterial and fungal surface macromolecules (Qureshi *et al.*, 1999a). The presence of pattern-recognition receptors in vertebrates, insects, and terrestrial plants has been demonstrated recently (O'Neill & Greene, 1998), and Toll-like receptors, involved in recognition of lipopolysaccharide (LPS, endotoxin) of gram-negative and lipoteichoic acid (LTA) of gram-positive bacterial pathogens, have been shown to play a key role in induced defense in these different groups of organisms (reviewed by Qureshi *et al.*, 1999b; Aderem & Ulevitch, 2000).

Despite the considerable knowledge about the role of surface macromolecules such as LPS and LTA in inflammatory reactions in humans, there are few investigations about their potential activities in organisms other than animals, in particular, in terrestrial plants and in algae. In mammalian inflammatory reactions, LPS and LTA are widely established as inducers of the synthesis of cytokines, nitric oxide (NO) and secondary messengers of the arachidonic acid cascade, such as prostaglandins by inducing enzymes of their synthesis such as cyclooxygenases (e.g. Chen *et al.*, 2001, Ichitani *et al.*, 2001). Arachidonic-acid derived prostaglandin E_2 is known to play a key role in protecting tissues against excessive damage by LPS-induced inflammatory reactions by regulating the intensities of the response (e.g. Jaworek *et al.*, 2001; Zhang & Rivest, 2001). In contrast, the vast majority of investigations concerning related structures in plants refer to lipo-chitooligosaccharidic Nod factures of rhizobial bacteria by their legume plant hosts prior to the establishment of symbiosis (e.g. reviewed by Cullimore *et al.*, 2001), with only few mentioning a role of LPS in the induction of defense mechanisms (e.g. Reitz *et al.*, 2000).

Prostaglandins in animals and jasmonate in plants are generated from arachidonic and linolenic acid, respectively. Interestingly, brown algae differ in their lipid profile from terrestrial plants due to their marked presence of eicosanoids, polyunsaturated fatty acids from the C_{20} family (e.g. Hofmann & Eichenberger, 1997). The presence of both arachidonic and linolenic acid in algae implies that both animal-like prostaglandins and plant-like cyclopentenones could be generated in these organisms in response to appropriate stimuli. Besides that, the presence of ROS could lead to the nonenzymatic formation of both C_{18}- and C_{20}-derived isprostanes, which have been described for animals and plants, respectively (Müller, 1998).

Populations of brown algae (Phaeophyceae) are plagued by various pathogens, including fungi and oomycetes (Küpper & Müller, 1999 / Chapter 2), plasmodiophoraleans (Maier *et al.*, 2000), viruses (Müller *et al.* 1998), bacteria (Sawabe *et al.*, 1998, 2000) and brown algal, pathogenic endophytes (Ellertsdóttir & Peters, 1997). Sporophytes of Laminariales, an order of brown algae with one of the most complex morphologies, have recently been shown to recognize fragments of their major cell polysaccharide, alginate, and to react with a rapid oxidative burst (Küpper *et al.*, 2001 / Chapter 6), a phenomenon discovered initially in human cells (Baldridge & Gerard, 1933). Signal transduction following elicitor recognition involves phospholipase A_2 (Küpper *et al.*, 2001 / Chapter 6), suggesting that free fatty acids play a role in the activation of defense responses in *Laminaria*, similar to higher plants (Chandra *et al.*, 1996).

Chemical defenses in marine brown algae include secondary metabolites such as terpenoids and polyphenols (Hay and Fenical, 1988, 1992; Steinberg, 1992; Hay and Steinberg, 1992; Cronin and Hay, 1996; Hay, 1996; Pavia and Toth, 2000), as well as fatty acid-derived compounds such as C_{11} hydrocarbons (Hay *et al.*, 1998; Schnitzler *et al.*, 1998; 2001). The latter are structural relatives of sexual pheromones of brown algae (Müller *et al.*, 1971; Maier, 1995). Like those, their biosynthesis originates from cleaved, polyunsaturated C_{20} fatty acids (Stratmann *et al.*, 1991; reviewed by Moore, 1999). The latter also seems to apply for C_{10} aldehydes playing a key role in the chemical defense of diatoms, a neighbouring phylum of brown algae (Miralto *et al.*, 1999). Apart from these, other studies concerning brown algal oxylipins, even though revealing a wealth of novel information, were descriptive in terms of structure and function, with only few functional aspects taken into account (Gerwick, 1994). Toth and Pavia (2000) recently provided evidence for the existence of water-borne cues, emitted by grazed individuals of *Ascophyllum nodosum*, and inducing defenses against the same grazers in neighboring individuals. Still, there is no knowledge as to the nature of molecules involved in distance and allelopathic signalling in brown algae and the marine environment in general. It is unclear whether compounds like jasmonic acid and derivatives have a role in the sea as they have in terrestrial ecosystems (Mur *et al.*, 1997), or whether other compounds fulfil a comparable role. Similarly, there is no information about the role of lipid-derived molecules in intracellular, defense-related signalling in brown algae.

The oxidative burst in *Laminaria* has been shown to play a crucial role to control the growth of epiphytic, potentially pathogenic bacteria (Küpper *et al.*, 2001 / Chapter 6; Chapter 9), producing hydrogen peroxide concentrations sufficient to significantly reduce bacterial growth and to reduce the number of microorganisms on algal surfaces. Furthermore, the oxidative burst is followed by an enhanced emission of iodinated halocarbons of higher microbial toxicity than the chlorinated or brominated compounds that are released in unstressed, steady-state conditions (Küpper, Kloareg & Malin, unpublished results). Still, it is not clear so far whether recognition of bacteria relies solely on endogenous elicitors such as oligoguluronates, released from the algal cell wall during bacterial attack, or whether brown algae can also recognize exogenous, bacterial elicitors: Whilst there is a certain body of knowledge on bacterial elicitors in higher plants (reviewed by Wojtaszek, 1997), there is only one report from a marine alga, the red agarophyte *Gracilaria conferta* (Weinberger & Friedlander, 2000).

We therefore challenged young *Laminaria digitata* thalli with a number of both lipoteichoic acid and lipopolysaccharide preparations from bacterial pathogens of humans and other mammals in order to draw first comparisons regarding recognition of conserved principles of eliciting molecules and a conserved eukaryotic defense response, the oxidative burst. Furthermore, these considerations induced us to search for oxidative responses to free fatty acids in *Laminaria digitata*.

Materials and Methods

Experimental cultures

Unialgal sporophytes were prepared by mating compatible pairs of gamethophytes: one pair of *Laminaria digitata* from Helgoland (German Bight), and two pairs of *Macrocystis pyrifera* from South America (Magellan Strait and Puerto Montt, Chile). The brown algal pathogenic endophyte, *Laminariocolax tomentosoides*, isolated from Helgoland, has been described previously (Burkhardt & Peters, 1998). Cultures were grown in Provasoli ES medium (Starr & Zeikus, 1987) prepared from artificial seawater (WIMEX, Krefeld, Germany, adjusted to 30 ‰ salinity), or natural Atlantic open-ocean water. They were illuminated with daylight-type fluorescent lamps at an irradiance of 9 $\mu E\ m^{-2}\ s^{-1}$ for 10 h per day and kept at 10 ±1°C. Cultures were transferred to fresh medium at one-or two-week intervals.

Chemicals

Arachidonic acid (5,8,11,14-eicosatetraenoic acid, C20:4), linolenic acid (9,12,15-octadecatrienoic acid, C18:3), linoleic acid (9,12-octadecadienoic acid, C18:2), oleic acid (9-octadecenoic acid, C18:1), and diphenylene iodonium (DPI) were obtained from SIGMA and dissolved in dimethyl sulfoxide (DMSO) to obtain stock solutions (100- or 1000-fold concentrated), depending on the experimental needs. Methyl jasmonate, pregnenolone, progesterone and clofibrate, were from SIGMA and were dissolved in DMSO to 3 mg. mL^{-1} for stock solutions; Radiolabeled ^{14}C-C20:4 and ^{14}C-C18:2 were from Amersham (Amersham, UK). Aspirin (acetylsalicylic acid), indomethacin and salicylhydroxamic acid (SHAM) were from SIGMA. Lipopolysaccharide (LPS) preparations from *Salmonella abortus-equi*, *Escherichia coli* (strain 0113 7/29/82), *Pseudomonas fluorescens* (wild type strain 10/2/87), *Salmonella minnesota* (strain 1114 8/30/70, LPS), *Klebsiella pneumoniae* (wild type strain 10/2/87, LPS 71), *Shigella dysenteriae* (wild type strain 8/22/80, LPS 77) originated from SIGMA.

Purified lipoteichoic acid preparations (LTA) were from *Bacillus subtilis* and *Staphylococcus aureus*, kindly provided by S. Morath, T. Hartung and A. Geyer, Universität Konstanz (Morath *et al.*, 2001).

Both LPS and LTA stock solutions were sonified prior to application to destroy micelles of these amphiphilic molecules, in order to render them accessible to potential receptors on the algal surface.

Hydrogen peroxide / ROS measurements

The concentration of hydrogen peroxide in the medium around algal thalli was determined using the luminol chemiluminescence method (Glazener *et al.*, 1991) with a LUMAT LB 9507 luminometer (EG&G Berthold, Bad Wildbach, Germany), as described by Küpper *et al.* (2001 / Chapter 6). Typically, 400 µL aliquots (out of 10 mL) were taken for one measurement. In the luminometer, 50 µL of 20 units mL^{-1} of horseradish peroxidase (Boehringer Mannheim, Meylan, France), dissolved in pH 7.8 phosphate buffer) and 100 µL of 0.3 M luminol (5-amino-2,3-dihydro-1,4-phthalazinedione, Sigma) solution were added to the sample. Chemiluminescence was recorded immediately after the last injection with a signal integration time of 10 s. For calculating the concentration of H_2O_2 present in the samples, calibration with a standard curve was carried out at least once during any series of measurements. Catalase at 100 U mL^{-1} and SOD at 200 U mL^{-1} (Boehringer Mannheim) were used to test the types of ROS detected in the assay.

TLC analysis

Laminaria digitata thalli (around 1 cm long; 1 – 1.5 g FW in 5 mL natural sea water) were incubated for 10, 60 and 120 min, respectively, in the presence of 5 µM ^{14}C-labelled arachidonic and linoleic acid, respectively. After the incubation time, a batch of thalli was extracted in 2 x 3 ml of ethyl acetate. The solvent was evaporated to a small volume and extracts were spotted on TLC plates, which were developed with a mixture of diethyl ether : light petroleum (boiling point, 40°C-60°C) : formic acid (50:50:1, v/v/v). The radioactivity of TLC plates was detected by means of an Instant-Imager (Packard-Instrument, USA).

LC-MS analysis

The metabolites derived from linoleic acid and arachidonic acid were resolved by RP-HPLC (Spectra-Physics Analytical) coupled to a Navigator LC/MS mass spectrometer (Finningan, Manchester, UK) equipped with an atmospheric pressure ionisation source (APCI) running on a negative ion mode (cone voltage 45 V).

The metabolites generated were resolved by RP-HPLC using a 5 µm Ultrasphere C18 column 150 x 4.6 mm (Beckman, France). The mobile phase (0.2 % acetic acid in water/acetonitrile) program began isocratically with a 40:60 mixture (v/v) for 35 min at a flow rate of 1 ml min^{-1}, followed by a 5 min-linear gradient to 5:95 (v/v) mixture for 20 min, in order to elute the residual substrate before returning to the initial conditions.

Infection experiments

To address the potential role of oxylipid signals in the induced resistance of kelps, three batches each of *L. digitata* sporophytes were elicited with either arachidonic acid or methyl jasmonate (0.5 μM) for 1 h, transferred to 50 ml fresh culture medium and inoculated at various intervals (1-8 days) after elicitation by *L. tomentosoides* zoospores, as follows. 500 μL of a culture of sporulating, free-floating filamentous endophytes were added to each dish of *Laminaria*. After co-culture for 3 days, the *Laminaria* sporophytes were transferred to new medium, which was then renewed weekly, and they were examined at three-day intervals until 3 weeks after the inoculation. Since a numerical scoring of infections was not feasible, the state of plants was judged by an unbiased, independent observer (Fig. 7). After the experiments, all plants were fixed, mounted on paper and deposited in the herbarium of the Marine Station, Roscoff.

Results

In a first step, compounds which are known to be involved in inflammatory reactions in mammals or defense of terrestrial plants were examined for their potential to trigger an oxidative burst in Laminariales. In a second step, two compounds were studied using a natural pathosystem (Chapter 9). Finally, preliminary analyses of the metabolic events involved with recognition of some of the compounds were carried out.

I. Oxidative burst

Effect of a peroxisome proliferator

Clofibrate (2-(p-chlorophenoxy-)2-methylpropionic acid ethyl ester), added at concentrations of 100 μM, triggered a strong oxidative burst in *L. digitata*. Whilst the response to oligoguluronates (150 μg. mL^{-1}) was immediate, the response to clofibrate resulted in a rather constant emission of hydrogen peroxide (Fig. 1).

Fatty acids and derivatives

Linolenic acid (C18:3) and arachidonic acid (C20:4) induced a response (Fig. 2 A), whilst oleic acid (C18:1) and linoleic acid (C18:2) were inactive. A dose-response curve revealed that the threshold concentration for arachidonic acid to trigger an oxidative burst lies between 0.1 and 1 μM. Higher concentrations (up to 200 μM, Fig. 2 B) were less active.

Cyclopentenone derivatives

Methyl jasmonate (MeJA) triggered a strong oxidative burst in *L. digitata* (Fig. 3 A), resembling the kinetics of the response to arachidonic acid. The threshold concentration was found at 1 μM and the strongest response observed was at 10 μM (Fig. 3 A). Similar to arachidonic acid, higher concentrations than 10 μM did not trigger stronger responses.

Among various prostaglandins (PGE_1, PGE_2, $PF_2\alpha$, PGB_1, PGA_2) tested, only PGA_2 was found to be active. It triggered the strongest response among all the compounds examined in this study (Fig. 3 B). The threshold concentration was found between 0.1 and 1 μM. In contrast to arachidonic acid and MeJA, the response increased in a dose-response-like manner towards higher concentrations, with the strongest response observed at 100 μM.

Steroids

Progesterone (500 μM, inactive at 10 μM) triggered an oxidative burst in one out of three *L. digitata* thalli tested, yet still with a fairly weak amplitude. In contrast, pregnenolone was inactive at both 10 and 500 μM.

Bacterial surface macromolecules: LPS and LTA

Lipoteichoic acid (LTA; Fig. 4 A) from *Bacillus subtilis* and lipopolysaccharide (LPS) from *Salmonella abortus-equi* (Fig. 4 B) are capable of inducing an oxidative burst in the brown alga *Laminaria digitata*. The threshold concentration for triggering a response was found between 1 and 10 μg. mL^{-1}. The initial response within the first 30 min after elicitation amounted to < 10 % and ca. 25 %, respectively, of the saturated H_2O_2 emission elicited by oligoguluronates. In contrast, these bacterial cell wall derivatives appeared to induce a second, much stronger oxidative burst starting around 30-60 min after their addition to the algae. Its amplitude reached or exceeded the level of maximum peroxide release triggered by GG. Lipoteichoic acid from *Staphylococcus aureus* (with and without purification) had no observable effect, and responses to 5 μg. mL^{-1} lipopolysaccharides from *Klebsiella pneumoniae* (LPS 71), *Escherichia coli* (LPS 63), and *Pseudomonas fluorescens* (LPS 70) were very weak.

Inhibitors

Diphenylene iodonium (DPI) at 10 μM inhibited the response to arachidonic acid, MeJA, and LPS from *Salmonella abortus-equi*, but it did not inhibit the response to PGA_2. Neither salicylhydroxamic acid (SHAM, a lipoxygenase inhibitor; Fabbri *et al.*, 2000), nor indomethacin (an inhibitor of cyclooxygenase; Jaworek *et al.*, 2001) had any significant effects on the oxidative burst triggered by arachidonic acid. Acetylsalicylic acid (aspirin), which inhibits both cyclooxygenase (Jaworek *et al.*, 2001) and allene oxide synthase (Pan *et al.*, 1998), had no inhibitory effect on the arachidonic acid-induced oxidative burst either. Propylgallate completely blocked the oxidative response in *Laminaria digitata*, being an effective scavenger of ROS.

Effects of arachidonic acid and methyl jasmonate on *Macrocystis pyrifera*

In order to extend the validity of the effects of arachidonic acid and methyl jasmonate to other brown algae, these two compounds were tested on *Macrocystis pyrifera* sporophytes. As shown in Fig. 5, both compounds induced a strong oxidative burst in this alga. Kinetics and amplitude were similar as in *L. digitata*, and the effect could also be inhibited by 10 μM DPI.

II. Analytics

L. digitata sporophytes were incubated for 10, 60 and 120 min with ^{14}C-labelled linoleic acid (which did not trigger an oxidative burst). Thin-layer chromatography of the extracted lipids and autoradiography showed that this compound was rapidly incorporated into the phospholipid fraction. No evidence for oxidized metabolites was found.

In contrast, ^{14}C-labelled arachidonic was rapidly metabolized, initiating between 10 and 60 min, as revealed by the disappearance of the band on the TLC sheet. Part of it was also traced in the phospholipid fraction. Subsequent LC-MS analysis of radiolabeled metabolites from the TLC plates pointed to the presence of derivatives with a retention time on RP-HPLC and a molecular mass expected for epoxides and hydroxylated derivatives of arachidonic acid (not shown).

III. Experimental pathosystem

Protection of *Laminaria digitata* against *Laminariocolax tomentosoides* by treatment with methyl jasmonate and arachidonic acid

The biological significance of elicitation by arachidonic acid and methyl jasmonate was addressed by infection experiments with young *L. digitata* sporophytes challenged with a natural, pathogenic endophyte of this species, *Laminariocolax tomentosoides* (Phaeophyceae). Non-elicited sporophytes were densely infested by endo- and epiphytic *Laminariocolax* filaments and had conspicuous necrotic lesions (Fig. 6 A-D). If young sporophytes were incubated for 2 hours with 0.5 μM of either arachidonic acid or MeJA, allowed to grow for 8 days in normal seawater medium and then challenged by *L. tomentosoides*, they were remarkably well protected against infection. Very few – if any - filaments of the endophyte were visible and even fewer necroses occurred (Fig. 6). Untreated *Laminaria* sporophytes showed a strong, localized accumulation of blue-autofluorescent compounds in the cortical cells around penetrating *Laminariocolax* filaments (Fig. 6 D), whilst the thalli pretreated with MeJA or arachidonic acid exhibited a less intense, but ubiquitous cortical accumulation of blue-autofluorescent compounds (Fig. 6 H and L, respectively).

As shown in Figure 7, only one day was required between elicitation with MeJA or arachidonic acid and inoculation with the endophyte to obtain a significant protection by the two compounds. The induced resistance persisted for 7 days.

Discussion

Increasing evidence points to the involvement of fatty acids and their oxidized derivatives in plant-pathogen interaction (Farmer, 1994; Farmer *et al.*, 1998). As shown by our study, brown algae seem to be no exception.

Similar to mammals, red and brown algae contain both arachidonic and linolenic acid at concentrations of 10 % or more of total fatty acids (Hoffmann & Eichenberger, 1997). Their oxidation and cyclisation to cyclopentenones, especially in the context of an oxidative burst (Küpper *et al.*, 2001 / Chapter 6), could yield both jasmonate- (C_{18}-derived) and prostaglandin-(C_{20}-derived)-like structures. Jasmonic acid and prostaglandins share structural features in being both cyclopentenones, resulting from the cyclisation of either linolenic or arachidonic acid.

The results presented here demonstrate that brown algae with complex thalli, such as *Laminaria digitata* and *Macrocystis pyrifera*, can recognize free linolenic and arachidonic acid as potent activators of defense reactions. Furthermore, this constitutes the first finding of free fatty acids triggering an oxidative burst, which is, to our knowledge, not known from terrestrial plants either. These molecules are likely to be released *in vivo* during pathogen attack, either directly by action of the pathogen on the cell membrane, or by phospholipase activation (Chandra *et al.*, 1996) in the signal transduction chain following elicitor recognition.

Furthermore, these results clearly demonstrate that brown algae recognize and produce cyclopentenones, including a prostaglandin A_2-like compound. Methyl jasmonate, however, has never been reported from marine algae so far, and this study yielded no evidence for its presence in brown algae either (data not shown).

Bostock *et al.* (1981) found that arachidonic acid can establish systemic resistance against *Phytophthora* in potato (*Solanum tuberosum*), attributing the recognition of arachidonic acid to its occurrence in the pathogen. In contrast, all other terrestrial plants investigated so far were insensitive to arachidonic acid. Methyl jasmonate is now well-known to be a volatile messengers between individuals of terrestrial plants (Mur *et al.*, 1997), where it induces resistance against pathogens, e.g. against *Pythium ultimum* in *Picea abies* (Kozlowski *et al.*, 1999). Our results showing the MeJA-induced resistance of *Laminaria* against *Laminariocolax* are the first experimental indication for the participation of cyclopentenones in distance signalling in the marine habitat, although the actual presence of this molecule under natural conditions is not yet confirmed. Nevertheless, recent evidence points to the existence of water-borne, resistance-inducing signalling compounds in brown algae (Toth & Pavia, 2000).

Our results obtained with steroids revealed only weak activities. Pregnenolone was completely inactive regarding induction of the oxidative burst, whilst progesterone was weakly active. A survey involving more derivatives alongside with an analytic approach is required to provide unequivocal evidence for the involvement of the steroid family in stress-related lipid signalling in Phaeophyceae.

There is an intriguing similarity between the effects of peroxisome proliferators in mammals and in plants. Clofibrate (2-(p-chlorophenoxy-)2-methylpropionic acid ethyl ester) induces cytochrome P450 enzymes catalyzing the ω-hydroxylation of fatty acids (Gibson *et al.*, 1982; Salaun *et al.*, 1986). The same compound has recently been demonstrated to stimulate the production of activated oxygen species and lipid peroxidation in pea seedlings, while catalase and superoxide dismutase were inhibited (Palma *et al.*, 91, Del Rio *et al.*, 96).

Peroxisome proliferators are known in mammals to trigger the release of free fatty acids and to induce enzyme activities which are involved in a cascade of fatty acid oxidation leading to a number of derivatives with potential bioactivity. Arachidonic acid itself and its oxidized derivative, prostaglandin J_2 (PGJ_2), are both ligands of nuclear peroxisome proliferator-activated receptors (PPAR; Kliewer *et al.*, 1995). Both of them should hence be able to induce expression of the same rapid and long-term responses. The presence of oxylipins including prostagladins (Bundy, 1985) in marine algae has been documented (Proteau & Gerwick, 1993; Rorrer *et al.*, 1997). Our results show clearly that both arachidonic and PGA_2 trigger a strong oxidative burst in *Laminaria.* Like in mammals (Kliewer *et al.*, 1995), arachidonic acid could be recognized by the same target receptors in *Laminaria* as its prostaglandin derivative: The absence of effects of the inhibitors (indomethacin, aspirin) of prostaglandin biosynthesis, cyclooxygenase (indomethacin and aspirin; Jaworek *et al.*, 2001), allene oxide synthase (aspirin; Pan *et al.*, 1998) and of lipoxygenase (SHAM) on the oxidative burst triggered by arachidonic acid may be considered as first indications in this aspect. But neither in terrestrial plants nor in algae, any evidence for PPAR-like receptors exists so far. On the other hand, the possibility that arachidonate and PGA_2 selectively activate the production of two independent sources of AOS could explain the lack of effect of inhibitors.

Whilst the oxidative burst triggered by arachidonic acid, MeJA and LPS can be inhibited by DPI, the response to PGA_2 can not. This finding indicates that at least two different sources of ROS co-operate in oxidative burst responses in brown algae. This is consistent with recent results in higher plants (e.g. Bolwell *et al.*, 1998; Bestwick *et al.*, 1999) which suggest that both an NAD(P)H oxidase (a neutrophil-like superoxide synthase, susceptible to inhibition by DPI, a suicide substrate of flavin-containing oxidases; O'Donnell *et al.*, 1993) and peroxidases should be involved the oxidative burst in higher plants.

Furthermore, it appears likely that the high concentrations of ROS generated during the oxidative burst could lead to the formation of isoprostanes from both linolenic and arachidonic acid. In animals, these nonenzymatically formed cyclopentanoic acid derivatives are known to greatly increase during oxidative stress and they are thought to have signalling functions in the responses to the stress (reviewed by Müller, 1998).

Up to now there have been no reports about lipoteichoic acids or lipopolysaccharides in the context of an oxidative burst in neither terrestrial plants nor any algal lineages. A few reports mention induction of systemic acquired resistance (e.g. Reitz *et al.*, 2000) by LPS in terrestrial plant-pathosystems, whilst others found a suppression of the hypersensitive reaction by bacterial LPS (Newman *et al.*, 1997). In no case investigated, LPS were found to trigger an oxidative burst in terrestrial plants (reviewed by Newman *et al.*, 2001; unpublished results mentioned by Dow *et al.*, 2000). Yet, there is a considerable body of literature concerning the structurally related chito-oligosaccharides mediating recognition between rhizobia and legume plants (reviewed by Cullimore *et al.*, 2001). Our study demonstrates that components of bacterial cell walls can be considered as exogenous elicitors in marine algae. Recognition of these surface structures pathogenic bacteria is not restricted to their natural mammalian hosts where they induce violent inflammatory responses, including an oxidative burst (Chen *et al.*, 2001; Ichitani *et al.*, 2001) , but that it occurs equally in a phylogenetically distant lineage, that separated from the ancestors of animals early in eukaryotic evolution. Since the bacteria which produce the examined LPS and LTA are not pathogenic to algae, these observations suggest that brown algae express unspecific receptors by which they can recognize common bacterial structures. It appears worthwile to search for the presence of Toll-like receptors in this context (Qureshi *et al.*, 1999b). LPS and LTA are interesting candidates to study enzymes and metabolites involved in the oxidative cascade of polyunsaturated fatty acids in brown algae: they are well-known to be potent triggers of eicosanoids in animal inflammatory reactions (e.g. Vafeas *et al.*, 1998). More specifically, among their multiple effects, they induce cyclooxygenase in arterial endothelium (Chen *et al.*, 2001), cytochrome P450-dependent fatty acid hydroxylases (CYP4A subfamily) and acyl-CoA oxidase in mammals (Barclay *et al.*, 1999).

The oxidative burst observed after challenging of *L. digitata* with arachidonic and linolenic acid, methyl jasmonate, LPS and LTA occurs considerably later than the response to oligosaccharides (Küpper *et al.*, 2001 / Chapter 6). It will have to be shown whether this is entirely due to physicochemical reasons such as the time needed to establish a distribution equilibrium of these mainly lipophilic compounds between the surrounding medium and the cell membrane, or if this lag phase is due to biochemical conversions of metabolites and induction of enzymes, or both.

In conclusion, the comparison between phyla as distant as mammals and brown algae, that separated during the so-called crown diversification of higher eukaryotes 600 million years ago (Bhattacharya *et al.*, 1991), appears worthwile, since it might allow to draw conclusions about conserved principles in the defense mechanisms of eukaryotes in general, regarding recognition of pathogens, signal transduction, and responses. The results of this study show that defense pathways exist in brown algae, which are similar to those known from animals and land plants. The features obviously conserved between such distant lineages include recognition of pathogenesis-related macromolecular patterns, eicosanoid and cyclopentenone signalling, and the capacity to produce an oxidative burst. These are probably evolutionary ancient features and we suggest that they have been a prerequiste for eukaryotic evolution.

Acknowledgements

We would like to thank Eva-Maria Boneberg and Siegfried Morath (University of Konstanz) for providing LPS and LTA preparations and stimulating discussions. Akira F. Peters (Institute for Marine Research, Kiel) provided cultures of *Laminariocolax tomentosoides*. Claire Gachon (Ecole Normale Supérieure, Paris) and Doris Bliestle (Photographic laboratory, University of Konstanz) contributed with microscopy and photography of infected plants, respectively. We would also like to acknowledge Georg Pohnert (Max Planck Institute for Chemical Ecology, Jena) for a critical reading of the manuscript and useful suggestions. F.C.K. was supported by fellowships from Studienstiftung des Deutschen Volkes (Bonn) and the European Commission (Program MAST-III).

References

Aderem A, Ulevitch RJ, 2000: Toll-like receptors in the induction of the innate immune response.- Nature 406, 782-787

Baldridge CW, Gerard RW, 1933: The extra respiration of phagocytosis.- Am. J. Physiol. 103, 235-236

Barclay TB, Peters JM, Sewer MB, Ferrari L, Gonzalez FJ, Morgan ET, 1999: Modulation of cytochrome P-450 gene expression in endotoxemic mice is tissue specific and peroxisome proliferator-activated receptor-alpha dependent.- J. Pharmacol. Exp. Ther. 290, 1250-1257

Bestwick C, Bolwell GP, Mansfield J, Nicole M, Wojtaszek P, 1999: Generation of the oxidative burst - scavenging for the truth.- Trends Pl. Sci. 4, 88-89

Bhattacharya D, Stickel SK, Sogin ML, 1991: Molecular phylogenetic analysis of actin genic regions from *Achlya bisexualis* (Oomycota) and *Costaria costata* (Chromophyta).- Journal of Molecular Evolution 33, 525-536

Blechert S, Brodschelm W, Holder S, Kammerer L, Kutchan T, Mueller M, Xia Z, Zenk M, 1995: The octadecanoid pathway: signal molecules for the regulation of secondary pathways.- Proc. Nat. Acad. Sci. USA 92, 4099-4105

Bolwell GP, Davies DR, Gerrish C, Auh C-K, Murphy TM, 1998: Comparative biochemistry of the oxidative burst produced by rose and French bean cells reveals two distinct mechanisms.- Plant Physiol. 116, 1379-1385

Bostock RM, Kuc JA, Laine RA, 1981: Eicosapentaenoic and arachidonic acids from *Phytophthora infestans* elicit fungitoxic sesquiterpenes in potato.- Science 212, 67-69

Bundy GL, 1985: Nonmammalian sources of eicosanoids.- Adv. Prostaglandin Thromboxane Leukot. Res. 14, 229-262

Burkhardt E, Peters AF, 1998: Molecular evidence from nrDNA IST sequences that *Laminariocolax* (Phaeophyceae, Ectocarpales *sensu lato*) is a worldwide clade of closely related kelp endophytes.- J. Phycol. 34, 682-691

Chandra S, Heinstein PF, Low PS, 1996: Activation of phospholipase A by plant defense elicitors.- Plant Physiol. 110, 979-986

Chen J-X, Berry L, Christman BW, Tanner M, Myers PR, Meyrick BO, 2001: NO regulates LPS-stimulated cyclooxygenase gene expression and activity in pulmonary artery endothelium.- Am. J. Physiol. Lung Cell Mol. Physiol. 280, L450-L457

Cronin G, Hay ME, 1996: Induction of seaweed chemical defenses by amphipod grazing.- Ecology 77, 2287-2301

Cullimore JV, Ranjeva R, Bono J-J, 2001: Perception of lipo-chitooligosaccharidic Nod factors in legumes.- Trends Pl. Sci. 6, 24-30

Del Rio LA, Palma JM, Sandalio LM, Corpas FJ, Pastori GM, Bueno P, Lopez-Huertas E 1996: Peroxisomes as a source of superoxide and hydrogen peroxide in stressed plants.- Biochem. Soc. Trans. 24, 434-438

Dow M, Newman MA, von Roepenack E, 2000: The induction and modulation of plant defense responses by bacterial lipopolysaccharides.- Ann. Rev. Phytopathol. 38, 1-21

Ellertsdóttir E, Peters AF; 1997: High prevalence of infection by endophytic brown algae in populations of *Laminaria* spp. (Phaeophyceae).- Mar. Ecol. Progr. Ser. 146, 135-143

Fabbri AA, Fanelli C, Reverberi M, Ricelli A, Camera E, Urbanelli S, Rossini A, Picardo M, Altamura MM, 2000: Early physiological and cytological events induced by wounding in potato tuber.- J. Exp. Bot. 51, 1267-1275

Farmer EE, Weber H, Vollenweider S, 1998: Fatty acid signaling in *Arabidopsis*.- Planta 206, 167-174

Farmer EE, 1994: Fatty acid signalling in plants and their associated microorganisms.- Plant Mol. Biol. 26, 1423-1437

Gerwick WH, 1994: Structure and biosynthesis of marine algal oxylipins.- Biochimica et Biophysica Acta 1211, 243-255

Gibson GG, Orton TC, Tamburini PP, 1982: Cytochrome P-450 induction by clofibrate.- Biochem. J. 203, 161-168

Glazener JA, Orlandi EW, Harmon GL, Baker CJ, 1991: An improved method for monitoring active oxygen in bacteria-treated suspension cells using luminol-dependent chemiluminescence.- Physiol. Mol. Plant Pathol. 39, 123-133

Hay ME, Piel J, Boland W, Schnitzler I, 1998: Seaweed sex pheromones and their degredation products frequently suppress amphipod feeding but rarely suppress sea urchin feeding.- Chemoecology 8, 91-98

Hay ME, 1996: Marine chemical ecology: what's known and what's next?- J. Exp. Mar. Biol. Ecol. 200, 103-134

Hay M, Steinberg PD, 1992: The chemical ecology of plant-herbivore interactions in marine versus terrestrial communities. In: Rosenthal, G.A., Berenbaum, M. R. (ed.): Herbivores: their interactions with secondary plant metabolites. Academic Press, Inc., 371-413

Hay ME, Fenical W, 1988: Marine plant-herbivore interactions: The ecology of chemical defense.- Ann. Rev. Ecol. Syst. 19, 111-145

Hofmann M, Eichenberger W, 1997: Lipid and fatty acid composition of the marine brown alga *Dictyopteris membranacea*.- Plant Cell Physiol. 38, 1046-1052

Ichitani Y, Holmberg K, Maunsbach AB, Haeggstrom JZ, Samuelsson B, De Witt D, Hokfelt T, 2001: Cyclooxygenase-1 and Cyclooxygenase-2 expression in rat kidney and adrenal gland after stimulation with systemic lipopolysaccharide: in situ hybridization and immunocytochemical studies.- Cell Tissue Res. 303, 235-252

Jaworek J, Bonior, Tomaszewska R, Jachimczak B, Kot M, Bielanski W, Pawlik WW, Sendur R, Stachura J, Konturek PC, Konturek SJ, 2001: Involvement of cyclooxygenase-derived prostaglandin E_2 and nitric oxide in the protection of rat pancreas afforded by low dose of lipopolysaccharide.- J. Physiol. Pharmacol. 52, 107-126

Kliewer SA, Lenhard JM, Wilson TM, Patel I, Morris DC, Lehmann JM, 1995: A prostaglandin J_2 metabolite binds peroxisome proliferator-activated receptor gamma and promotes adipocyte differentiation.- Cell 83, 813-819

Kozlowski G, Buchala A, Métraux J-P, 1999: Methyl jasmonate protects Norway spruce (*Picea abies* (L.) KARST) seedlings against *Pythium ultimum* TROW.- Physiol. Mol. Plant Pathol. 55, 53-58

Küpper FC, Müller DG, 1999: Massive occurrence of the heterokont and fungal parasites *Anisolpidium*, *Eurychasma* and *Chytridium* in *Pylaiella littoralis* (Ectocarpales, Phaeophyceae).- Nova Hedwigia 69, 381-389 (Chapter 2 of this thesis)

Küpper FC, Kloareg B, Guern J, Potin, P, 2001: Oligoguluronates elicit an oxidative burst in the brown algal kelp *Laminaria digitata*.- Plant Physiol. 125, 278-291 (Chapter 6 of this thesis)

Lien E, Sellati TJ, Yoshimura A, Flo TH, Rawadi G, Finberg RW, Carroll JD, Espevik T, Ingalls RR, Radolf JD, Golenbock DT, 1999: Toll-like receptor 2 functions as a pattern recognition receptor for diverse bacterial products. J. Biol. Chem. 274, 33419-33425

Maier I, Parodi E, Westermeier R, Müller DG, 2000: *Maullinia ectocarpii* gen. et sp. nov. (Plasmodiophorea), an intracellular parasite in *Ectocarpus siliculosus* (Ectocarpales, Phaeophyceae) and other filamentous brown algae.- Protist 151, 225-238

Maier I, 1995: Brown algal pheromones.- Chapter in: Progress in Phycological Research 11 (Round FE and Chapman DJ, eds), 51-102. Biopress Ltd., Bristol

Miralto A, Barone G, Romano G, Poulet SA, Ianora A, Russo GL, Buttino I, Mazzarella G, Laabir M, Cabrini M, Giacobbe MG, 1999: The insidious effect of diatoms on copepod reproduction.- Nature 402, 173-176

Moore BS, 1999: Biosynthesis of marine natural products: microorganisms and macroalgae.- Nat. Prod. Reports 16, 653-674

Morath S, Geyer A, Hartung T, 2001: Structure-function relationship of cytokine induction by lipoteichoic acid from *Staphylococcus aureus*.- J. Exp. Med. 193, 393-398

Müller DG, Kapp M, Knippers R, 1998: Viruses in marine brown algae.- Adv. Virus Res. 50, 49-67

Müller DG, Jaenicke L, Donike M, Akintobi T, 1971: Sex attractant in a brown alga: chemical structure.- Science 171, 815-817

Müller MJ, 1998: Radically novel prostaglandins in animals and plants: the isoprostanes.- Chemistry & Biology 5, R323-R333

Mur LAJ, Kenton P, Draper J, 1997: Something in the air: volatile signals in plant defence.- Trends Microbiol. 5, 297-300

Newman M-A, Dow JM, Daniels MJ, 2001: Bacterial lipopolysaccharides and plant-pathogen-interactions.- Eur. J. Pl. Pathol. 107, 95-102

Newman M-A, Daniels MJ, Dow JM, 2001: The activity of Lipid A and core components of bacterial lipopolysaccharides in the prevention of the hypersensitive response in pepper.- Mol. Plant-Microbe Interact. 10, 926-928

O'Donnell V, Tew DG, Jones OTG, England PJ, 1993: Studies on the inhibitory mechanism of iodonium compounds with special reference to neutrophil NADPH oxidase.- Biochem. J. 290, 41-49

O'Neill LA, Greene C, 1998: Signal transduction pathways activated by the IL-1 receptor family: ancient signaling machinery in mammals, insects, and plants.- Leukoc. Biol. 63, 650-657

Orozco-Cárdenas ML, Narváez-Vásquez J, Ryan CA, 2001: Hydrogen peroxide acts as a second messenger for the induction of defense genes in tomato plants in response to wounding, systemin, and methyl jasmonate.- Plant Cell 13, 179-191

Palma JM, Garrido M, Rodriguez-Garcia MI, Del Rio LA, 1991: Peroxisome proliferation and oxidative stress mediated by activated oxygen species in plant peroxisomes.- Arch. Biochem. Biophys. 287, 68-74

Pan Z, Camara B, Gardner HW, Backhaus RA, 1998: Aspirin inhibition and acetylation of the plant cytochrome P450, allene oxide synthase, resembles that of animal prostaglandin endoperoxide H synthase.- J. Biol. Chem. 273, 18139-18145

Pavia H, Toth G, 2000: Inducible chemical resistance in the brown seaweed *Ascophyllum nodosum*.- Ecology 81, 3212-3225

Proteau PJ, Gerwick WH, 1993: Divinyl ethers and hydroxy fatty acids from three species of *Laminaria* (brown algae).- Lipids 28, 783-787

Pugin J, Heumann ID, Tomasz A, Kravchenko VV, Akamatsu Y, Nishijima M, Glauser MP, Tobias PS, Ulevitch RJ, 1994: CD14 is a pattern recognition receptor.- Immunity 1, 509-516

Qureshi ST, Lariviere L, Leveque G, Clermont S, Moore KJ, Gros P, Malo D, 1999a: Endotoxin-tolerant mice have mutations in Toll-like receptor 4 (Tlr4).- J. Exp. Med. 189, 615-625

Qureshi ST, Gros P, Malo D, 1999b: Host resistance to infection. Genetic control of lipopolysaccharide responsiveness by TOLL-like receptor genes.- Trends Genet. 15, 291-294

Reitz M, Rudolph K, Schröder I, Hoffmann-Hergarten S, Hallmann J, Sikora RA, 2000: Lipopolysaccharides of *Rhizobium etli* Strain G12 act in potato roots as an inducing agent of systemic resistance to infection by the cyst nematode *Globodera pallida*.- Appl. Env. Microbiol. 66, 3515-3518

Reymond P, Farmer EE, 1998: Jasmonate and salicylate as global signals for defense gene expression.- Curr. Opin. Plant Biol. 1, 404-411

Rorrer G, Yoo H, Huang Y, Hayden C, Gerwick W, 1997: Production of hydroxy fatty-acids by cell-suspension cultures of the marine brown alga *Laminaria saccharina*.- Phytochemistry 46, 871-877

Salaun J-P, Simon A, Durst F, 1986: Specific induction of lauric acid ω-hydroxylase by clofibrate, diethylhexyl-phtalate and 2,4-dichlorophenoxyacetic acid in higher plants.- Lipids 21, 776-779

Sawabe T, Tanaka R, Iqbal MM, Tajima K, Ezura Y, Ivanova EP, Christen R, 2000: Assignment of *Alteromonas elyakovii* KMM 162^T and five strains from spot-wounded fronds of *Laminaria japonica* to *Pseudoalteromonas elyakovii* comb. nov. and the extended description of the species.- Int. J. Syst. Evol. Microbiol. 50, 265-271

Sawabe T, Makino H, Tatsumi M, Nakano K, Tajima K, Iqbal MM, Yumoto I, Ezura Y, Christen R, 1998: *Pseudoalteromonas bacteriolytica* sp. nov., a marine bacterium that is the causative agent of red spot disease in *Laminaria japonica*.- Int. J. Systematic Bacteriol. 48, 769-774

Schnitzler I, Pohnert G, Hay ME, Boland W, 2001: Chemical defense of brown algae (*Dictyopteris* spp.) against the herbivorous amphipod *Ampithoe longimana*.- Oecologia 126, 515-521

Schnitzler I, Boland W, Hay ME, 1998: Organic sulfur compounds from Dictyopteris spp. deter feeding by an herbivorous amphipod (*Ampithoe longimana*) but not by an herbivorous sea urchin (*Arbacia punctulata*).- J. Chem. Ecol. 24, 1715-1732

Starr RC, Zeikus JA, 1987: UTEX - the culture collection of algae at the University of Texas at Austin.- J. Phycol. 23 (Suppl.), 1-47

Steinberg PD, 1992: Geographical variations in the interaction between marine herbivores and brown algal secondary metabolites. In: V.J. Paul (ed.): Ecological roles of marine natural products. Cornell University, New York, USA, 51-92

Stratmann K, Boland W, Müller DG, 1992: Pheromone mariner Braunalgen, ein neuer Zweig des Eicosanoidstoffwechsels.- Angew. Chem. 104, 1261-1263

Toth GB, Pavia H, 2000: Water-borne cues induce chemical defense in a marine alga (*Ascophyllum nodosum*).- Proc. Nat. Acad. Sci. USA 97, 14418-14420

Vafeas C, Mieyal PA, Urbano F, Falck JR, Chauhan K, Berman M, Schwartzman ML, 1998: Hypoxia stimulates the synthesis of cytochrome P450-derived inflammatory eicosanoids in rabbit corneal epithelium.- J. Pharmacol. Exp. Ther. 287, 903-910

Weinberger F, Friedlander M, 2000: Endogenous and exogenous elicitors of a hypersensitive response in *Gracilaria conferta* (Rhodophyta).- J. Appl. Phycol. 12, 139-145

Weinberger F, Friedlander M, 2000: Endogenous and exogenous elicitors of a hypersensitive response in *Gracilaria conferta* (Rhodophyta).- J. Appl. Phycol. 12, 139-145

Wojtaszek P, 1997: Oxidative burst: an early plant response to pathogen infection.- Biochem. J. 322, 681–692

Zhang J, Rivest S, 2001: Anti-inflammatory effects of prostaglandin E_2 in the central nervous system in response to brain injury and circulating lipopolysaccharide.- J. Neurochemistry 76, 855-864

Figures

Fig. 1. Peroxisome proliferator. Induction of an oxidative burst by clofibrate (100 μM) in *Laminaria digitata*. Compared to elicitation by GG (Küpper *et al.*, 2001 / Chapter 6), the response takes place only after a 45- to 60-min latency, but with a comparable amplitude. Addition of clofibrate is marked by an arrow.

Fig. 2. Free fatty acids. A) Linolenic acid (C18:3, 100 μM, triangles) triggers an acid oxidative burst in *Laminaria digitata*, whilst linoleic acid (C18:2, 100 μM, squares) and oleic acid (not shown) do not (for comparison: effect of oligoguluronates, 100 μg ml^{-1}, diamonds). **B)** Dose-response behavior of the oxidative burst triggered by arachidonic acid in *Laminaria digitata* (Diamonds: 100 μM, circles: 10 μM, squares: 1 μM, asterisks: 0.1 μM). The threshold concentration for arachidonic acid to trigger an oxidative burst in *L. digitata* is between 0.1 and 1 μM.

Fig. 3. Cyclopentenones. A, B) Dose-response behaviour of the oxidative burst triggered by prostaglandin A_2 in *L. digitata*: 100 μM (-), 50 μM (+), 10 μM (•), 6 μM (∗), 3 μM (X), 1 μM (△), 0.1 μM (□), control (◇). **C)** Dose-response behaviour of the oxidative burst triggered by methyl jasmonate in *L. digitata*: Diamonds: 100 μM, circles: 10 μM, squares: 1 μM, asterisks: 0.1 μM. The active threshold concentration is at 1 μM, and the strongest response observed at 10 μM. Addition of compounds marked is by an arrow.

Fig. 4. Bacterial endotoxins. A) Lipoteichoic acid from *Bacillus subtilis* (10 μg ml^{-1}, circles) trigger an oxidative burst in *L. digitata*, whilst LTA from *Staphylococcus aureus* (10 μg ml^{-1}, filled triangles; 1 μg ml^{-1}, hollow triangles) does not (for comparison: control, diamonds; 150 μg ml^{-1} GG, squares). **B)** Lipopolysaccharide from *Salmonella abortus-equi* (10 μg ml^{-1}, filled circles; 1 μg ml^{-1}, hollow circles) triggers an oxidative burst in *L. digitata* (for comparison: control, diamonds; 150 μg ml^{-1} GG, squares). Addition of compounds is marked by an arrow.

Fig. 5. Arachidonic acid **(A)** and methyl jasmonate **(B)** trigger an oxidative burst in *Macrocystis pyrifera*, which can be inhibited by 10 μM DPI (diamonds: 200 μM arachidonic acid or methyl jasmonate, respectively; squares: same addition of arachidonic acid or methyl jasmonate, respectively, but pre-treated for 20 min with 10 µM DPI). Addition of compounds is marked by an arrow.

Fig. 6. Results of infection / protection experiments, with *Laminaria digitata* sporophytes challenged by the pathogenic endophyte *Laminariocolax tomentosoides*.
A-D) Untreated control. A, B) Surface of *L. digitata*, completely covered by *L. tomentosoides* filaments (scale bars: 1 mm - A, 1 cm - B). C, D) Section of a strongly infected thallus from the same batch, in bright field (C) and UV-excitation (D). Penetrating filaments are visible by their reddish chlorophyll autofluorescence in the largely chloroplast-free medulla. The zone of penetration is typically surrounded by an accumulation of bright-blue autofluorescent compounds (scale bars C, D: 10 µm).
Protection against *Laminariocolax* infection by treatment with either arachidonic acid (E-H) or methyl jasmonate (J-M), one day prior to inoculation with the pathogen. The surface of treated plantlets (E-F, J-K) is free of any visible symptom of infection, confirmed by bight-field (G, K) and UV-fluorescence (H, L) microscopy, and by marks of an independent observer (I, M). Treated plants often show a diffuse cortical accumulation of bright-blue autofluorescent compounds (H, L). Scale bars: E, G, H, J, L, M: 10 µm; F, K: 1 cm.

Fig. 7. Time scale of resistance induction by arachidonic acid (A) and methyl jasmonate (B) against *L. tomentosoides* in *L. digitata*. The graph shows the temporal establishment and the level of C20:4- and MeJA-primed resistance against *L. tomentosoides* in *L. digitata*, based on the marks given by an independent observer (error bars indicate variability, expressed as the standard deviation of the marks attributed to a given series of plants).The extent of damage inferred by infections was judged according to the following relative scale: 0) massive infections, all plants dead; 1) strong infections, at least one plant of the batch dead; 2) strong infections and necroses in several plants; 3) strong infections or necroses in at least one plant; other plants of the same batch in good shape; 4) few infections or necroses: overall plants in good shape; 5) infections and / or necroses only very minor, 6) no symptoms of pathogen infection nor unusual necroses (other than the usual senescence at the apical part of the blade).

Figure 1

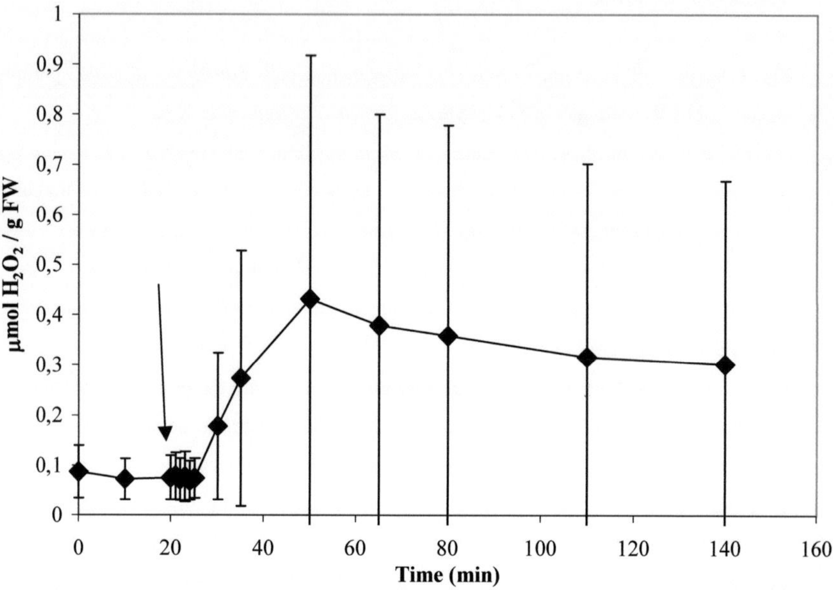
µmol H_2O_2 / g FW
1
0,9
0,8
0,7
0,6
0,5
0,4
0,3
0,2
0,1
0
0
20
40
60
80
100
120
140
160
Time (min)

Figure 2

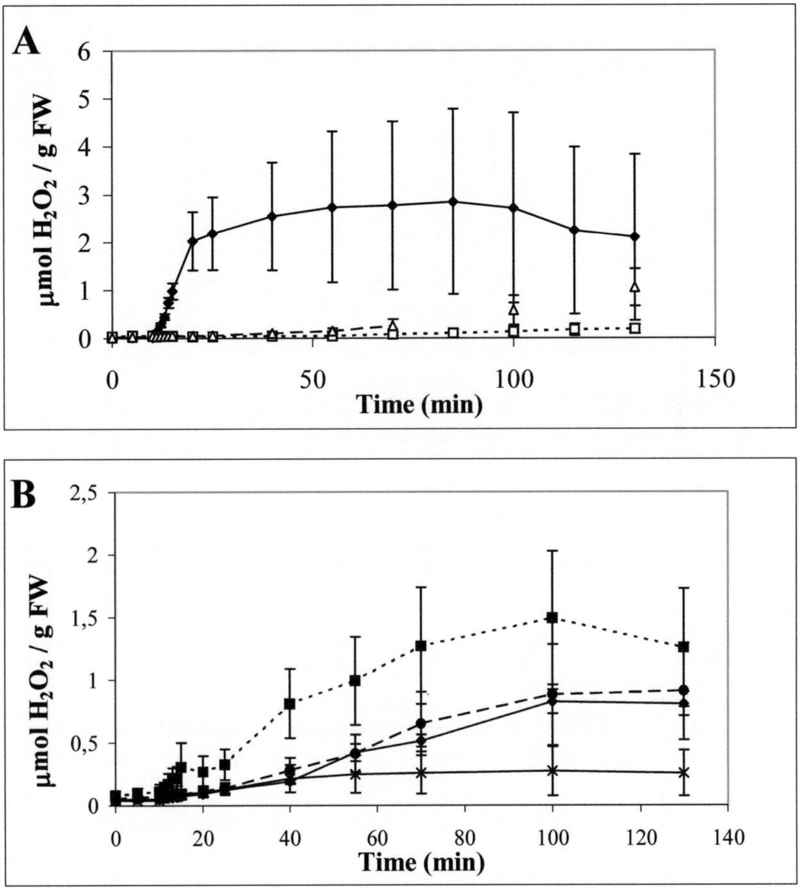
A
6
5
4
3
2
1
0
μmol H2O2 / g FW
0
50
100
150
Time (min)
B
2,5
2
1,5
1
0,5
0
μmol H2O2 / g FW
0
20
40
60
80
100
120
140
Time (min)

Figure 3

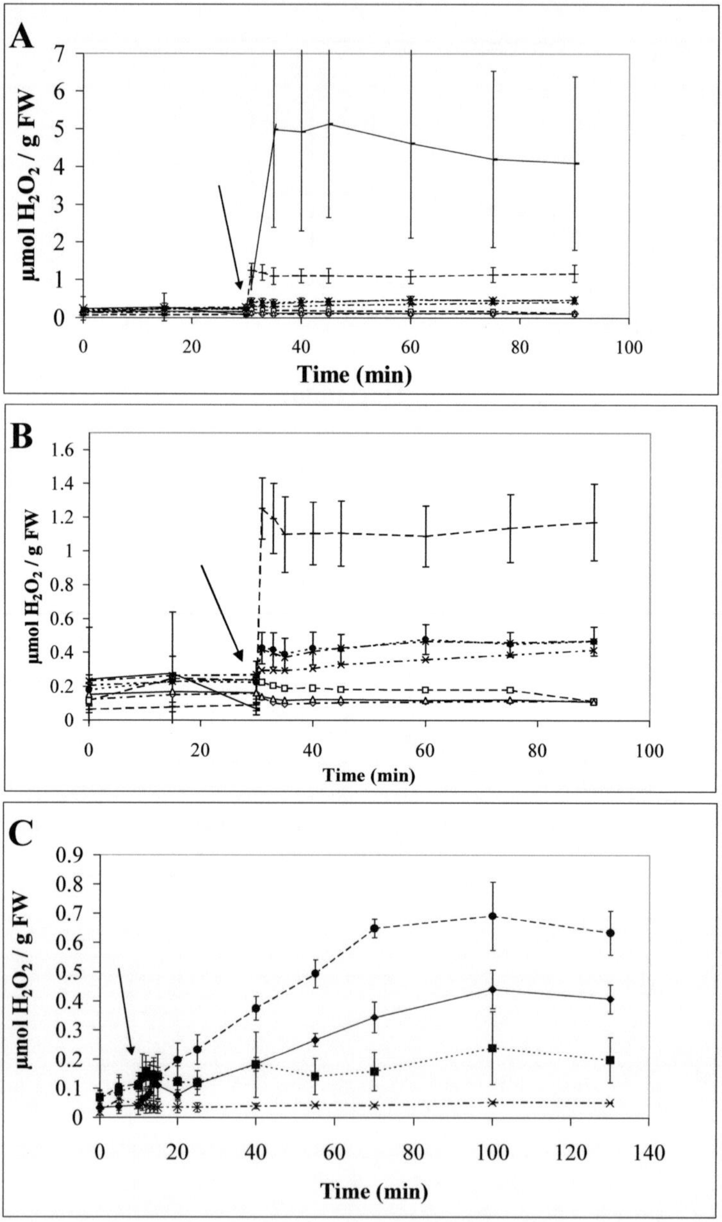
A
µmol H2O2 / g FW
7
6
5
4
3
2
1
0
0
20
40
60
80
100
Time (min)
B
µmol H2O2 / g FW
1.6
1.4
1.2
1
0.8
0.6
0.4
0.2
0
0
20
40
60
80
100
Time (min)
C
µmol H2O2 / g FW
0.9
0.8
0.7
0.6
0.5
0.4
0.3
0.2
0.1
0
0
20
40
60
80
100
120
140
Time (min)

Figure 4

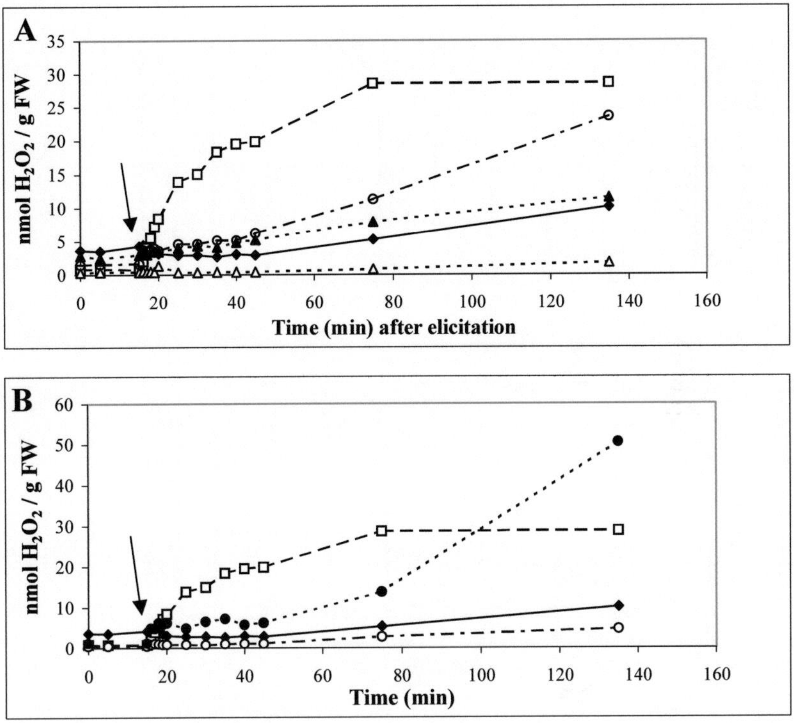
A
35
30
25
20
15
10
5
0
nmol H_2O_2 / g FW
0
20
40
60
80
100
120
140
160
Time (min) after elicitation
B
60
50
40
30
20
10
0
nmol H_2O_2 / g FW
0
20
40
60
80
100
120
140
160
Time (min)

Figure 5

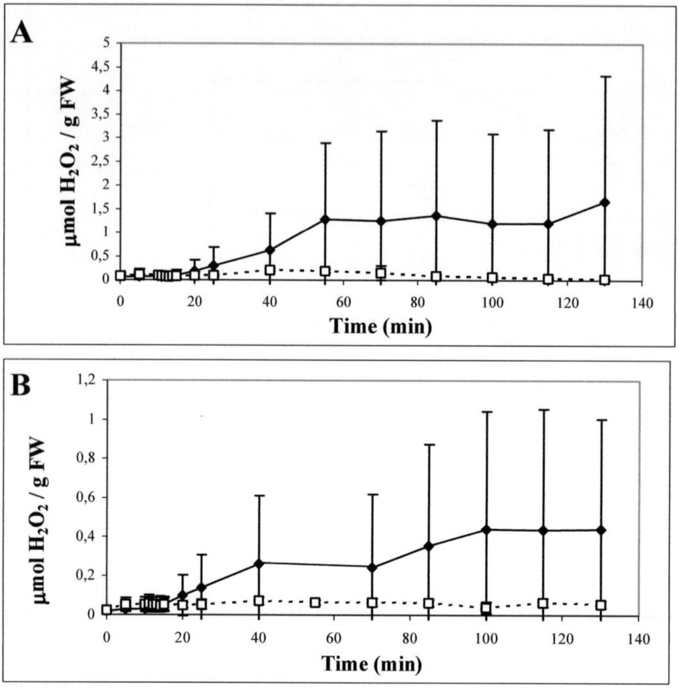
A
µmol H2O2 / g FW
5
4,5
4
3,5
3
2,5
2
1,5
1
0,5
0
0
20
40
60
80
100
120
140
Time (min)
B
µmol H2O2 / g FW
1,2
1
0,8
0,6
0,4
0,2
0
0
20
40
60
80
100
120
140
Time (min)

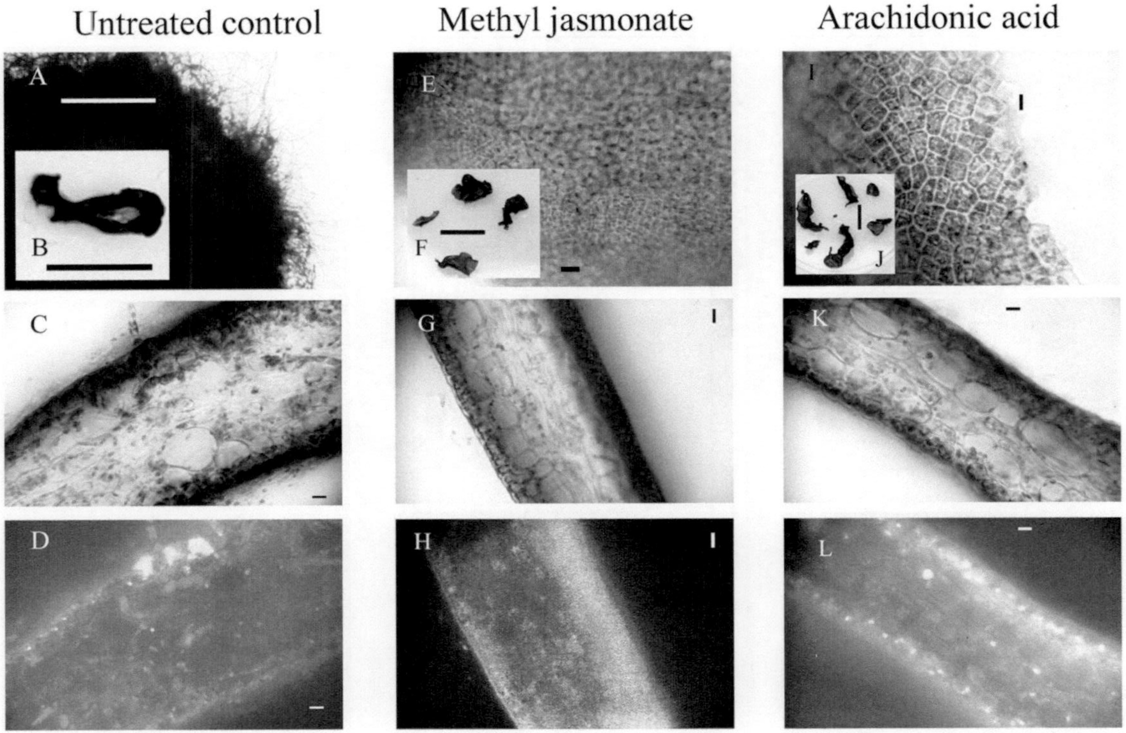
Untreated control
Methyl jasmonate
Arachidonic acid
A
B
C
D
E
F
G
H
I
J
K
L

Figure 6

Figure 7

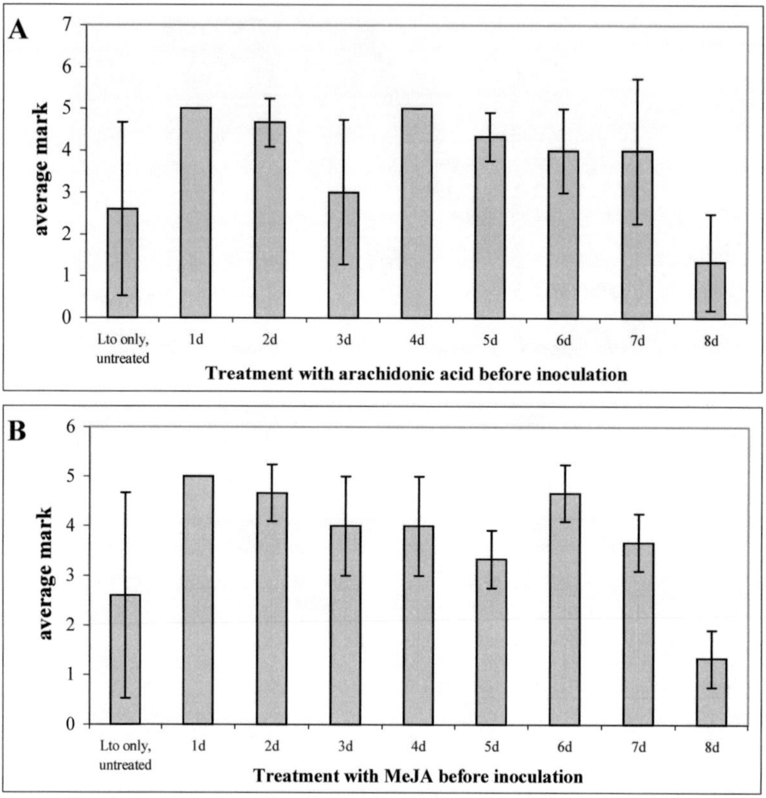
A
average mark
7
6
5
4
3
2
1
0
Lto only, untreated
1d
2d
3d
4d
5d
6d
7d
8d
Treatment with arachidonic acid before inoculation
B
average mark
6
5
4
3
2
1
0
Lto only, untreated
1d
2d
3d
4d
5d
6d
7d
8d
Treatment with MeJA before inoculation

11

General discussion and outlook:

Phycopathology on the rise – interdisciplinary approaches to decipher host-pathogen interactions in brown algae

Ecology of pathogens and parasites in natural brown algal populations

The lack of knowledge concerning the role of pathogens and pests in natural systems is a major gap in understanding population and ecosystem processes (Real, 1996). Brown algal populations are plagued by various pathogens, including prokaryotes, eukaryotes and viruses, and the recent years have seen an increase in research activity in this field.

Ectocarpales populations show high prevalences of virus infections (Sengco *et al.*, 1996; Müller *et al.*, 2000), but eukaryotic pathogens are also commonplace in this group: Wilce *et al.* (1982) observed frequent, seasonal epidemics of *Eurychasma dicksonii* in planktonic *Pylaiella littoralis* off the coast of Massachusetts, contributing to vegetative propagation by breaking of filaments at the infection sites. New field observations within the present study (Chapter 1) demonstrated that *Eurychasma dicksonii*, *Anisolpidium rosenvingei* and *Chytridium polysiphoniae*, two heterokonts and a fungus, frequently occur epidemically in populations of *Pylaiella littoralis* over a wide stretch of the European Atlantic coast. The *Anisolpidium-Pylaiella* interaction is a case in which host fertility rather than host biomass constitutes the limiting factor for spread of the pathogen in a prey-predator-like interaction (Fig. 1). An analogous case of reproduction-linked pathology was described by Martínez and Correa (1993) for the kelps *Lessonia nigrescens* and *L. trabeculata*, with an epiphyte (*Sporocladopsis* sp.) specifically affecting the sori.

A cross-infection survey (Chapter 2) demonstrated that *E. dicksonii* (and to a lesser extent, *C. polysiphoniae*) infects members of many orders of brown algae, confirming and extending the already wide host range based on field observations (e.g. Sparrow, 1960; Jenneborg, 1977). Given the numerous records from temperate and cold localities all over the northern hemisphere (Sparrow, 1960) and a few southern hemisphere localities (e.g. Ricker, 1987), *Eurychasma* has to be considered as the most widespread marine plant pathogen. The temperature optima observed in laboratory culture experiments agree with the cold-temperate

Fig. 1. Hypothetical sequence of events during the seasonal cycle of the *Pylaiella littoralis* – *Anisolpidium rosenvingei* interaction (Chapter 1). **A**: presence of *Anisolpidium rosenvingii* infections. **Pr**: reproductive period of *Pylaiella littoralis*. **Pb**: biomass of *Pylaiella littoralis*. Dotted lines: not visible in field habitat.

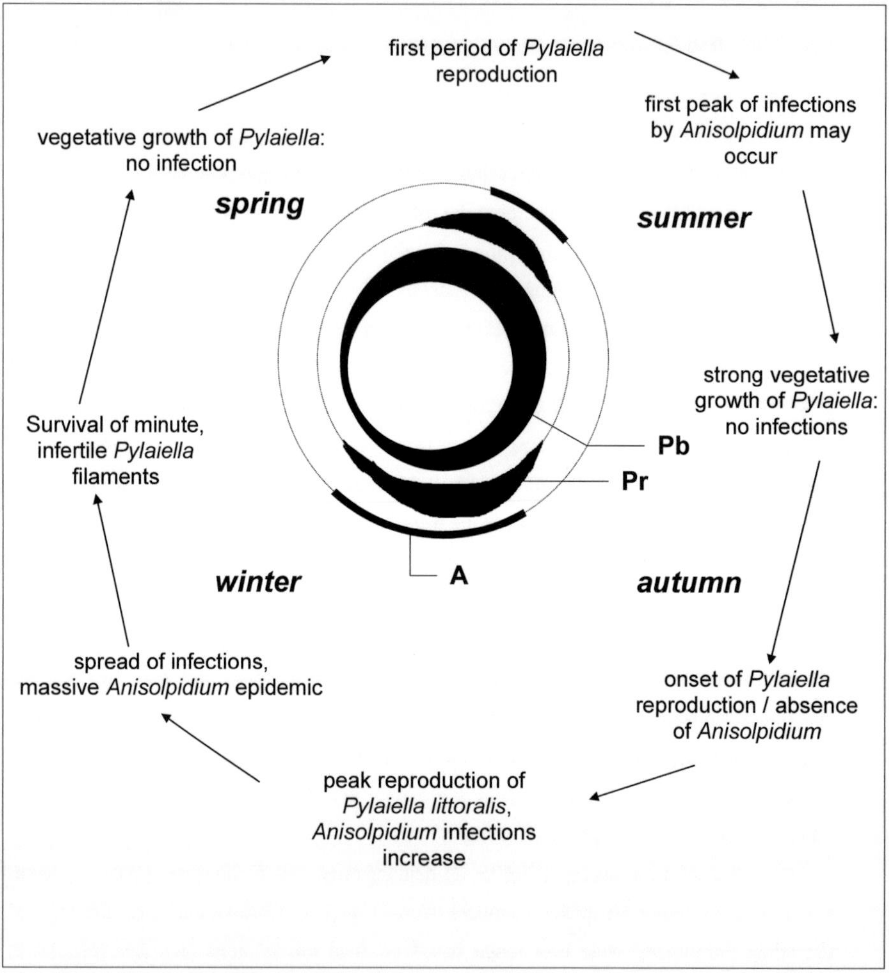

distribution of *Eurychasma* and *Chytridium* (Chapter 2). With regard to its wide host range including most brown and probably also a number of red algae (e.g. Sparrow, 1960), it appears possible that *E. dicksonii* may act as a vector for gene transfer between brown algal species, and possibly even between different lineages. Brown algal viruses offer the possibility to study triangular interactions between a brown algal host, a virus and a fungal or heterokont parasite. Preliminary experiments have shown that the presence of a virus infection in a gametophyte of *Ectocarpus siliculosus* renders this algal strain resistant against infection by *P. polysiphoniae* (D.G. Müller, *personal communication*), suggesting that acquired resistance may also exist in brown algae.

Maier *et al*. (2000) found a novel plasmodiophoralean pathogen in Chilean *Ectocarpus siliculosus*. In culture it infected members of other brown algal orders, including gametophytes of Laminariales. Infection symptoms are inconspicuous, suggesting that this parasite may be widespread but frequently overlooked. It may be concluded that careful search will detect other novel parasites.

In an ultrastructural study, Pellegrini and Pellegrini (1982) suggested that brown algal surfaces are colonized and lysed by the combined action of various morphologically distinguishable bacteria. Nevertheless, most older reports concerning bacterial infections remain doubtful as far as causality is concerned (reviewed by Andrews, 1976), with the exception of two recent cases in Japanese mariculture systems of *Laminaria japonica*, in which the causative agents were identified: spot-wounded disease, caused by *Pseudoalteromonas elyakovii* (Sawabe *et al*., 2000), and red-spot disease with the pathogen *Pseudoalteromonas bacteriolytica* (Yumoto *et al*., 1989; Sawabe *et al*., 1998).

Kelp sporophytes are frequently infested by Phaeophycean endophytes which cause thallus alterations (Apt, 1988; Kawai and Tokuyama, 1995; Peters & Schaffelke, 1996; Ellertsdóttir & Peters, 1997; Burkhardt & Peters, 1998; Fig. 2) probably decreasing resistance of affected plants against drag (Peters, personal communication). Ellertsdóttir and Peters (1997) reported high prevalences of endophytic brown algae in kelp populations at Helgoland (German Bight), but this phenomenon is probably more widespread.

Phylogeny of pathogens affecting marine algae

Marine plant pathogens comprise several eukaryotic lineages as well as bacteria and viruses. Within the last 5 years, molecular methods have been used to clarify phylogenetic affiliations among some groups. Most commonly, the sequences of the small subunit (SSU,

Fig. 2. Endophytes are widespread in natural kelp populations, such as this infection of *Laminaria digitata* by *Streblonema deformans* in Chenal de l'Ile de Batz, Roscoff, Brittany (the twisted part of the stipe is ca. 0.8 m long; underwater photograph by the author, 1997).

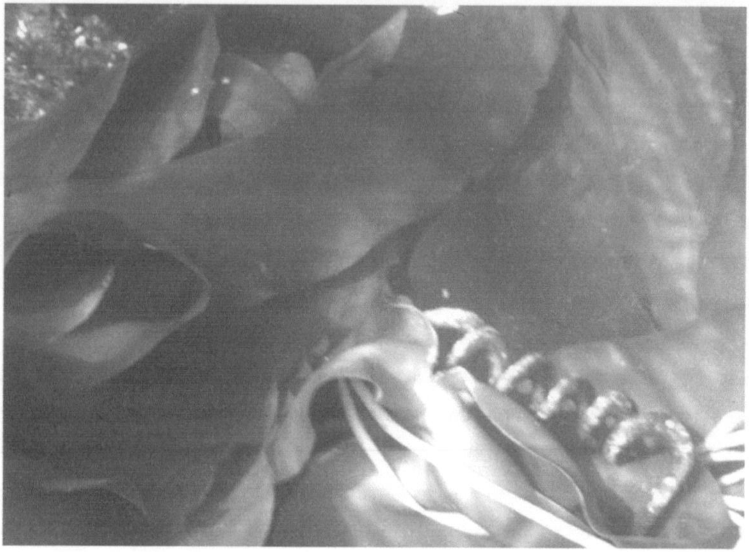

18 S) ribosomal RNA gene and the internal transcribed spacer (ITS) of the ribosomal cistron were used. Whilst the SSU rRNA is a marker for phylogenetic distances between entire phyla above ordinal level (Saunders & Kraft, 1995; Tan & Druehl, 1996), the faster-evolving ITS is useful at species and genus level (Saunders & Druehl, 1993; Peters *et al.*, 1997; Stache-Crain *et al.*, 1997).

Regarding brown algal endophytes of Laminariales, Peters & Burkhardt (1998) showed that *Laminarionema elsbetiae* belongs to the Ectocarpales *sensu lato*, but that it is not affiliated to any of the major subgroups. Burkhardt and Peters (1998) provided evidence that *Laminariocolax* is a genus comprising pathogens from the kelp genera *Laminaria*, *Ecklonia* and *Macrocystis*, which are distributed worldwide in cold and temperate seas.

After disputes concerning their phylogenetic position dating back over a century, the present study (Chapter 4) provided clear answers by studying the 18 S rRNA sequence homologies of *Eurychasma dicksonii* and *Chytridium polysiphoniae*. The results show that *Eurychasma* belongs to the Oomycota, branching at an ancestral position between terrestrial plant pathogens and free-living marine heterotrophic picoplankters. *Chytridium polysiphoniae*, a true fungus, appears more closely related to the genera *Rhizophydium* and *Spizellomyces* (Chytridiomycota) than to other *Chytridium* species, calling for a taxonomic rearrangement.

Similarly, Sawabe *et al.* (1998; 2000) clarified the taxonomic position of two bacterial pathogens of kelps recently, using 16 S rDNA sequence homologies (*Pseudoalteromonas bacteriolytica* and *Pseudoalteromonas elyakovii*, respectively).

Viruses infecting brown algae have been frequently described in the past (for reviews see Van Etten *et al.*, 1991; Reisser, 1993). More recent molecular studies provided more information for a comprehensive understanding, including phylogenetic aspects (Müller *et al.*, 1998). Delaroque *et al.* (2001) provide the first complete sequence of a marine virus. Pringle (1998) suggested to place the known brown algal viruses as a new genus *Phaeovirus* in the family of Phycodnaviridae. This is justified by their special features like a circular genome structure, a complex infection cycle, a non-reactivity with primers directed against α-like DNA polymerases, extended latency, and the well-balanced relationship with their hosts.

Metabolism and bioenergetics of host-pathogen interactions in marine algae

Parasitism and symbiosis can be considered as opposite extremes of a metabolic interaction between a host and an associated organism, but co-evolution often results in an initially parasitic relationship developing towards symbiosis. In this context, bioenergetic aspects have a key role. Goff (1979, 1982) reported on translocation of photoassimilates from red algal hosts to parasitic red algae with little or no photosynthetic pigments, whilst Penot (1974) demonstrated the transfer of ^{14}C-glutamate and various radio-labelled elements between *Ascophyllum nodosum* and the red algal epiphyte *Polysiphonia lanosa*.

The availability of *E. dicksonii* and *C. polysiphoniae* in laboratory cultures opens a new range of perspectives for future study. Overall, the interaction of *E. dicksonii* with its host appears much more temperate than that of *C. polysiphoniae*, suggesting a more advanced stage of co-evolution. Whilst *C. polysiphoniae* soon kills the host cells, as demonstrated by the rapid accumulation of UV-autofluorescent compounds and cessation of photosynthesis, *E. dicksonii* stimulates infected host cells to hypertrophic growth. Concomitant stimulation of host photosynthesis during the first half of the infection cycle (Chapter 4) probably plays a key role. Except for a recent study (Chou *et al.*, 2000), the underlying mechanisms of exogenous control of plastid photosynthesis by biotrophic pathogens are still largely unknown to date. An analogous situation of pathological hypertrophic growth led to the discovery of the gibberellins as a class of higher plant hormones, produced by the fungal pathogen *Fusarium moniliforme* (= *Gibberella fujikuroi*; reviewed by Phinney, 1983). The association of *Eurychasma* with brown algal hosts may be a candidate for a search of brown algal growth hormones. Despite reports about occurrence and effects of compounds known as hormones from higher plants in macroalgae (e.g. Bradley, 1991), evidence for an actual *in vivo* function remains scarce. During the first half of its infection cycle, the intracellular existence of *Eurychasma* in a host cell appears more like another cell organelle than that of a pathogen. This could be viewed as an early phase of a pathway leading to endosymbiosis. Furthermore, the growing interest in natural products from novel sources, in particular, marine organisms and microorganisms (e.g. Pettit, 1996; Liberra & Lindequist, 1995), may use marine Oomycota and Chytridiomycota with these two species as interesting, potential new sources.

Current understanding of defense pathways in *Laminaria*

Among brown algae, *Laminaria digitata* is presently the most intensely studied example regarding defense pathways. Experimental evidence and working hypotheses from various parts of this complex are summarized in Fig. 3. Under steady-state conditions, an apoplastic vanadium haloperoxidase (***1***) oxidizes iodide from the seawater, yielding hypoiodous acid (HIO). This compound either crosses the cell membrane directly (as an unpolar molecule, undissociated at the pH of sea water) or after halogenating an organic molecule (***2***). In the cytosol, rapid reduction takes place (***3***), effectively creating a concentration gradient from the outside to the inside of the cell. An important question still to be resolved is the precise intracellular location and chemical form of accumulated iodine (***4***), which can reach levels as high as 5 % dry weight in young *Laminaria* plants.-

Fig. 3. Scheme summarizing the current knowledge of inducible defense in Laminariales sporophytes

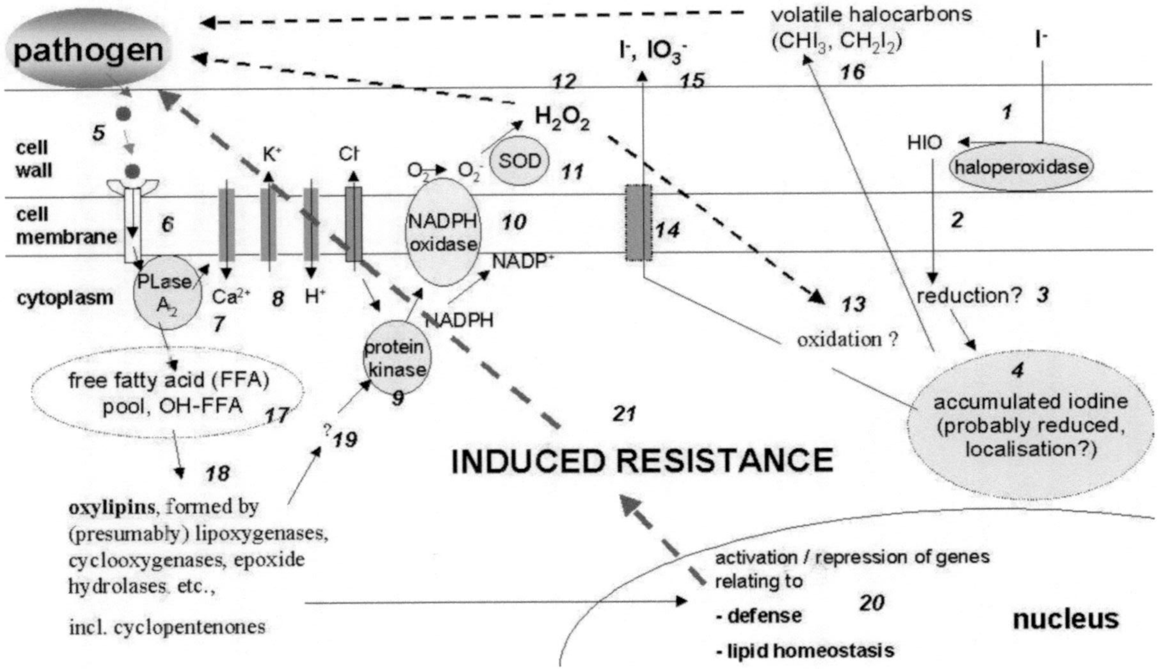

In case of an attack by an alginolytic pathogen – which could be a bacterium, a fungus, an oomycete, or a pathogenic brown algal endophyte such as *Laminariocolax* – the brown algal cell wall is enzymatically degraded, releasing various oligomeric degradation products of carbohydrates (***5***). Among these, oligoguluronates are recognized by a putative receptor at the surface of the host cell (***6***), triggering a signal transduction chain: Phospholipase A_2 (***7***) hydrolyses membrane lipids. Its action yields free fatty acids and activates further elements, including ion channels (***8***) and protein kinases (***9***), finally turning on a superoxide-producing, presumably NADPH-consuming oxidase (***10***). This enzyme consumes molecular oxygen (measurable as a respiratory burst), yielding superoxide, which is rapidly converted by an apoplastic superoxide dismutase (SOD) (***11***), yielding hydrogen peroxide in the apoplast (***12***). This has a direct cytotoxic effect on the attacking pathogen outside the cell. Furthermore, being a less polar molecule than water, it can more easily cross membranes, i.e. enter the cytosol, where it has probably stress-related signalling functions. Upon that, it contributes to the mobilization of the reservoir of accumulated iodine (***13***): A single oxidative burst involving the whole surface of a young *Laminaria* plant can liberate over 10 % of the accumulated iodine, which leaves the cell by a still unknown mechanism (***14***) most of which is encountered in the surrounding medium in the form of iodide and iodate, i.e. partly oxidized (***15***). Besides that, the oxidative burst is followed by an increased emission of iodinated halocarbons (***16***), which may be considered as detoxification products of hydrogen peroxide, but which also have a higher toxicity than brominated or chlorinated halocarbons, suggesting that iodine accumulation in *Laminaria* has both anti-oxidant and defense functions. Free fatty acids (***17***) are also potent activators of defense reactions in *Laminaria*. Physiologically, they are released following elicitation by the action of phospholipases such as PLA_2 (***7***). They can undergo various oxidation reactions (***18***), yielding oxylipins, highly active signalling molecules. Among them are cyclopentenones such as prostaglandins and possibly jasmonic acid derivatives. As a result, the superoxide-producing machinery is further activated directly by a still unknown mechanism (***19***), and genes related to defense and lipid homoiostases in the nucleus are upregulated (***20***) – resulting in the establishment of induced resistance against further pathogen attack (***21***). The latter is probably a multi-component process including enhanced cell wall crosslinking and expression of genes producing defense compounds, including at least one new haloperoxidase isoform.

Halogen metabolism of marine algae in the context of stress and defense

The present study has established a link between reactive oxygen species (ROS) in the apoplasm, V-haloperoxidase and iodine uptake in *Laminaria digitata* (Chapter 5). Furthermore, these results show that the iodine reservoir of the cells can be mobilized upon oxidative stress, consistent with observations of an increased emission of volatile halogenated compounds (VHCs; Collén *et al.*, 1994; Mtolera *et al.*, 1996; Pedersén *et al.*, 1996). These findings indicate that iodine may have redox-regulatory functions in brown algae. A similar function of iodine was suggested for animals, in the context of both evolution of terrestrial life and the epidemiology of thyroid cancer (Venturi *et al.*, 2000). The consistently observed deficit between the amount of radioactive label taken up from the medium and the amount detected in the thalli (usually ranging between 10 and 30 %) suggested that a considerable part of the metabolized iodine is volatilized, probably in the form of VHCs (e.g. Carpenter *et al.*, 2000). Still, important remaining questions concern the intracellular localization and chemical state of accumulated iodine, and the mechanism of its rapid efflux. The intracellular chemistry could be addressed by extended X-ray absorbtion fine structure (EXAFS) and X-ray absorption near-edge structure (XANES) spectroscopy, two techniques that have been successfully applied for sulfur-accumulating bacteria (Prange *et al.*, 1999). Furthermore, energy-dispersive analysis of X-rays (EDAX) appears suitable for addressing intracellular iodine localization. This technique has been used in metal-hyperaccumulating terrestrial plants to unveil the precise localization of accumulated heavy metals (e.g. Küpper *et al.*, 1999). Another important aspect is the biosynthetic pathway of volatile halogenated, in particular, iodinated compounds. Bromoperoxidase seems to be involved in one pathway (Theiler *et al.*, 1978; Wever *et al.*, 1991), and methyl chloride transferase has been suggested to be involved in the formation of chlorinated methanes in a halotolerant higher plant, a red alga and a fungus (Wuosmaa and Hager, 1990) by another pathway. In my view, it is imaginable that VHCs are derived from halogenated degradation products of lipid peroxidation in the cell membrane. The availability of high concentrations of ROS, reactive I_2, Br_2, HIO and HBrO in close proximity to unsaturated fatty acids, readily susceptible to halogenation by electrophilic addition, support this hypothesis. In addition, recent findings point at iodinated lipids (Küpper and Eichenberger, preliminary results) and the abundance of VHCs with hydrocarbon chains of different length, C_1, C_2 and fractions of longer compounds (Chapter 8).

Defense reactions in marine algae share some conserved features with mammals and terrestrial plants

The present study has demonstrated that *Laminaria* sporophytes recognize oligoalginates as endogenous elicitors, leading to a strong, transient oxidative burst (Chapter 6). This finding in brown algae, which represent a lineage very distant from higher plants, suggests that elicitation and the oxidative burst are ubiquitous phenomena among living organisms and a phylogenetically old feature. The burst observed in *L. digitata*, a highly evolved brown alga, is in many aspects reminiscent of defense reactions observed in higher plants. Similarities are found in elicitors, the time course, the indications for the signal transduction pathways involving protein phosphorylation, ion channels, superoxide as the first generated reactive oxygen species, and desensitisation after a first elicitation.

Lipid oxidation products – oxylipins - from seaweeds have encountered a growing interest in recent years. Although a number of novel compounds were described, research remained rather descriptive and oriented towards potential pharmaceutical applications (Gerwick *et al.*, 1993). Pheromones of brown algae have been intensely studied, and information from biosynthesis to *in vivo* function is available (for review: Maier, 1995). Similarly, Schnitzler *et al.* (1997) describe sulfur-C_{11}-compounds with defense functions in the brown seaweed *Dictyopteris*. Chapter 8 of this study provided further information on functional aspects of lipids. Various lipid derivatives were found to be potent triggers of defense reactions in kelps: lipoteichoic acid (LTA), lipopolysaccharides (LPS) from bacterial pathogens of humans (where ample knowledge in their analogous role in inflammatory reactions exists), polyunsaturated fatty acids such as arachidonic acid and linolenic acid, a compound known as a volatile stress messenger in higher plants, methyl jasmonate, and finally, prostaglandin A_2 (PGA_2), which produced the strongest reactions (Chapter 8).

The oxidative burst in *Laminaria* is followed by a rapid efflux of various iodine compounds, which may amount to almost 20 % of the total accumulated iodine in one single elicited plant (Chapter 7). These conditions – high concentrations of ROS, halogenides, presence of haloperoxidases –likely favour the generation of halogenating oxidants, another interesting analogy to the oxidative burst in human eosinophils (Weiss *et al.*, 1986). As highly electrophilic agents hypohalous acids can react with a variety of organic compounds and they are considered as potent natural biocides (Wever et al., 1991). Furthermore, the emission of iodinated halocarbons is strongly enhanced after elicitation, whilst the inorganic species comprise mainly iodide (I^-) and iodate (IO_3^-). Toxicity trials with a kelp-associated bacterium showed that the growth-inhibitory effect of a given halogenated compound increases in the series Cl < Br < I, and that it depends on the number of halogen atoms in a molecule, favouring multiple-iodinated compounds over brominated and chlorinated ones (Küpper and Potin, unpublished). These observations are consistent with results obtained with various halocarbons against the pathogenic endophyte of the red alga *Chondrus crispus*, *Acrochaete operculata* (Bouarab & Potin, personal communication). Overall, our results with iodine metabolism in *Laminaria* suggest that it has both anti-oxidant and defense functions.

Ecological implications of defense reactions of marine algae

The biological significance of the oligoalginate-triggered oxidative burst is obvious from experiments with endophytic kelp pathogens (*Laminariocolax* sp.) and alginolytic, epiphytic bacteria (Chapter 9). *Laminariocolax* endophytes have a high peroxide-scavenging capacity, suggesting that they can suppress the oxidative burst of their host upon infection, which may partly explain their large success in natural kelp populations (Ellertsdóttir & Peters, 1996). Experimental inhibition of the superoxide source involved in the oxidative burst by diphenylene iodonium makes kelp plants highly susceptible to attack by epiphytic, alginolytic bacteria that are normally moderate commensals. This resembles recent results in the red agarophyte *Gracilaria conferta* by Weinberger and Friedlander (2000). Furthermore, this study provides the first conclusive evidence for the induction of efficient resistance mechanisms against pathogens in brown algae.

In analogy to certain higher plants, arachidonic acid (Bostock *et al.*, 1981) and methyl jasmonate (e.g. Kozlowski *et al.*, 1999; for review: Mur *et al.*, 1997), respectively, mediate a rapid, induced resistance against *Laminariocolax*.

The occurrence of a rapid and strong oligoalginate-triggered oxidative burst is largely restricted to alginate-rich, morphologically complex sporophytes of the Laminariales. Members of most other orders of brown algae (e.g. *Ectocarpus* sp., Küpper & Müller, preliminary results) tend to release considerable quantities of peroxide and phenolic exudates constitutively. Some of the latter have been suggested to have anti-herbivore and allelopathic properties (Wolfe, 1988). In the Desmarestiales, only one (*Desmarestia dudresnayi*) out of three species tested produced an oxidative burst. Although equipped with a similarly complex morphology as the Laminariales, this group has developed sulfuric acid accumulation as an alternative defense strategy (McClintock *et al.*, 1982). Interestingly, the host spectrum of

Eurychasma (Chapter 3) is largely inverse to the occurrence of a rapid and strong oligoalginate-triggered oxidative burst among brown algae.

The comparison of *L. digitata* as one of the most complex and *P. littoralis* as a simple-structured brown alga on the one side, and of *Laminariocolax* and *Eurychasma* on the other, offers the opportunity to address the evolution of defense systems and mechanisms to control ROS among brown algae. *L. digitata* sporophytes, like other species of the Laminariales and Fucales, accumulate high levels of iodine and bromine, contain V-haloperoxidases (Chapter 5), exhibit a high peroxide scavenging capacity and are capable to react to oligoalginate elicitors with a rapid, strong oxidative burst (Chapter 6). This is followed by a massive release of potentially toxic iodinated compounds and can be seen as a first step to induced resistance. *P. littoralis* as a primitive filamentous form reacts to elicitation with a rather weak oxidative burst, does not accumulate halogens, has no detectable haloperoxidase activity (Vilter *et al.*, 1983), but continuously excretes large amounts of ROS and maintains a high concentration of H_2O_2 in its surrounding. In most of these aspects, it resembles gametophytes, the haploid, microscopic life form of kelps. It is legitimate to conclude that, compared to *L. digitata*, *Pylaiella* is a rather primitive system not only on the structural and developmental level, but also with regard to defense and to ROS metabolism. The capability to trigger an oxidative burst and to effectively control ROS levels both on a temporal and spatial scale may be considered as the essential prerequisites for the development towards more complex structures in the phylum Phaeophyta.

What's next?

The continuing, fast advance of genomics is likely to contribute to the understanding of algal host-pathogen interactions. The application of an expressed sequence tag (EST) approach (Crépineau et al., 2000) within the framework of a larger genome project on marine model organisms (Boyen, Potin, Kloareg, *personal communication*) to compare elicited and non-elicited *Laminaria digitata* and *Chondrus crispus* plants is likely to identify defense and resistance-related genes. Availability of DNA microarrays will allow to monitor expression of a multitude of genes in the contexts of stress, defense and induced resistance (Reymond, 2001). Knowledge of their sequences should allow to draw conclusions concerning the phylogeny of defense mechanisms within a larger context of eukaryotic evolution. The respiratory burst oxidase, previously cloned from humans (Royer-Pokora *et al.*, 1986), rice

(Groom *et al.*, 1996), *Arabidopsis* (Torres et al., 1998), tomato, and *Potamogeton* (Amicucci *et al.*, 1999), would be of particular interest, since it is a key enzyme of a defense mechanism that is considerably more ancestral than adaptive mammalian immunity based on antibody-antigen-recognition. The availability of the first complete sequence of a marine virus (Delaroque *et al.*, 2001) will foster better understanding of the molecular bases of algal physiology, including defense, and to reveal first clues about the existence of lateral gene transfer in marine algae. Transformation of heterokont organisms (Falciatore *et al.*, 1999) will open similar perspectives of studying functions of defense reactions as in higher plants, e.g. using transgenic tobacco expressing the *Pseudomonas putida* salicylate hydroxylase gene (Gaffney *et al.*, 1993; Delaney *et al.*, 1996).

Neutral molecular markers such as microsatellites recently became available (Billot *et al.*, 1998) for monitoring population structure in these algae. On the other hand, the identification of defense genes in the host species (see above) will provide tools for monitoring the population structure of genes involved with defenses. The comparison of the spatial distribution and diversity of neutral genetic markers *versus* that of these selected genes is of primary interest for understanding the dynamics of host-parasite evolution and, more specifically, the efficiency of local adaptation.

Following first evidence for distance signalling in brown algae (Toth & Pavia, 2000), chemical ecology now faces the challenge of identifying the molecules involved.

In conclusion, we now for the first time have the possibility of experimentally mimicking attacks by pathogens in marine algae and of exploring the nature and the transduction pathways of the induced defenses. Establishing whether the key points underlined above actually are involved in chemical defense against pathogens and grazers can be now studied by elicitation with oligosaccharides, oxylipins and other yet unidentified elicitors from algal pathogens, followed by monitoring the induced responses and by assays of acquired resistance. Combining investigations in both laboratory cultures and in the field will give information on the threshold concentrations for defensive activity and on the biotope levels under stressed or normal conditions and contribute major insights in the regulation mechanisms that control the environmental level of these metabolites. Systematic investigation of gene expression in the presence or absence of elicitor signals will identify genes involved in these defenses. When biochemical or gene defense markers are available, the search will be open for waterborne or airborne (in intertidal species), distance communication molecules.

References

Amicucci E, Gaschler K, Ward JM, 1999: NADPH oxidase genes from tomato (*Lycopersicon esculentum*) and curly-leaf pondweed (*Potamogeton crispus*).- Plant Biology 1, 524-528

Andrews JH, 1976: The pathology of marine algae.- Biological Reviews 51, 211-253

Apt KE, 1988: Etiology and development of hyperplasia induced by *Streblonema* sp. (Phaeophyta) on members of the Laminariales (Phaeophyta).- Journal of Phycology 24, 28-34

Billot C, Rousvoal S, Estoup A, Epplen JT, Saumitou-Laprade P, Valero M, Kloareg B, 1998: Isolation and characterization of microsatellite markers in the nuclear genome of the brown alga *Laminaria digitata* (Phaeophyceae).- Molecular Ecology 7, 1771-1788

Bostock RM, Kuc JA, Laine RA, 1981: Eicosapentaenoic and arachidonic acids from *Phytophthora infestans* elicit fungitoxic sesquiterpenes in potato.- Science 212, 67-69

Bradley PM, 1991: Plant hormones do have a role in controlling growth and development of algae. Journal of Phycology 27, 317-321

Burkhardt E, Peters AF, 1998: Molecular evidence from nrDNA sequences that *Laminariocolax* (Phaeophyceae, Ectocarpales sensu lato) is a worldwide clade of closely related kelp endophytes.- Journal of Phycology 34, 682-691

Carpenter L.J., Malin G., Küpper F.C., Liss P.S., 2000: Novel biogenic iodine-containing trihalomethanes and other short-lived halocarbons in the coastal East Atlantic.- Global Biogeochemical Cycles 14, 1191-1204

Chou H-M, Bundock N, Rolfe SA, Scholes JD, 2000: Infection of *Arabidopsis thaliana* leaves with *Albugo candida* (white blister rust) causes a reprogramming of host metabolism.- Molecular Plant Pathology 1, 99-113

Collén J, Ekdahl A, Abrahamsson K, Pedersén M, 1994: The involvement of hydrogen peroxide in the production of volatile halogenated compounds by *Meristiella gelidium* (Rhodophyta).- Phytochemistry 36, 1197-1202

Crépineau F, Roscoe T, Kaas R, Kloareg B, Boyen C, 2000: Characterisation of complementary DNAs from the expressed sequence tag analysis of life cycle stages of *Laminaria digitata* (Phaeophyceae).- Plant Molecular Biology 43, 503-513

Delaney TP, Uknes S, Vernooij B, Friedrich L, Weymann K, Negrotto D, Gaffney T, Gut-Rella M, Kessmann H, Ward E, Ryals J, 1994: A central role of salicylic acid in plant disease resistance.- Science 266, 1247-1250

Delaroque N, Müller DG, Bothe G, Pohl T, Knippers R, Boland W, 2001: The complete DNA sequence of the Ectocarpus siliculosus virus EsV-1 genome.- Virology, *in press*

Ellertsdóttir E, Peters AF, 1997: High prevalence of infection by endophytic brown algae in populations of *Laminaria* spp. (Phaeophyceae).- Marine Ecology Progress Series 146, 135-143

Falciatore A, Casotti R, Leblanc C, Abrescia C, Bowler C, 1999: Transformation of a non-selectable reporter gene in marine diatoms.- Marine Biotechnology 1, 239-251

Gaffney T, Friedrich L, Vernooij B, Negrotto D, Nye G, Uknes S, Ward E, Kessmann, Ryals J, 1993: Requirement of salicylic acid for the induction of systemic acquired resistance.- Science 261, 754-756

Gerwick WH, 1994: Structure and biosynthesis of marine algal oxylipins.- Biochimica et Biophysica Acta 1211, 243-255

Goff LJ, 1979: The biology of *Harveyella mirabilis* (Cryptonemiales, Rhodophyceae). VI. Translocation of photoassimilated ^{14}C.- Journal of Phycology 15, 82-87

Goff LJ, 1982: The biology of parasitic red algae.- Chapter 8 in: Progress in Phycological Research 1 (Round FE and Chapman DJ, eds), 289-369. Elsevier Biomedical Press B.V.

Groom QJ, Torres MA, Fordham-Skelton AO, Hammond-Kosack KE, Robinson NJ, Jones JDG, 1996: *rbohA*, a rice homologue of the mammalian *gp91phox* respiratory burst oxidase gene.- Plant Journal 10, 515-522

Jenneborg L-H, 1977: *Eurychasma*-infection of marine algae. Changes in algal morphology and taxonomical consequences.- Botanica Marina 20: 499-507

Kawai H, Tokuyama M, 1995: *Laminarionema elsbetiae* n. gen., n. sp. (Ectocarpales, Phaeophyceae), a new endophyte in *Laminaria* sporophytes.- Phycological Research 43, 185-190

Kozlowski G, Buchala A, Métraux J-P, 1999: Methyl jasmonate protects Norway spruce (*Picea abies* (L.) Karst) seedlings against *Pythium ultimum* Trow.- Physiological and Molecular Plant Pathology 55, 53-58

Küpper H, Zhao F, McGrath, 1999: Cellular compartmentation of zinc in leaves of the hyperaccumulator *Thlaspi caerulescens*.- Plant Physiology 119, 305-311

Liberra K, Lindequist U, 1995: Marine fungi – a prolific source of biologically active natural products?- Pharmazie 50, 583-587

Maier I, Parodi E, Westermeier R, Müller DG, 2000: *Maullinia ectocarpii* gen. et sp. nov. (Plasmodiophorea), an intracellular parasite in *Ectocarpus siliculosus* (Ectocarpales, Phaeophyceae) and other filamentous brown algae.- Protist 151, 225-238

Maier I, 1995: Brown algal pheromones.- in: Progress in Phycological Research 11 (Round FE and Chapman DJ, eds), 51-102. Biopress Ltd., Bristol

Martínez E, Correa J, 1993: Sorus-specific epiphytism affecting the kelps *Lessonia nigrescens* and *L. trabeculata*.- Marine Ecology Progress Series 96, 83-92

McClintock M, Higinbotham N, Uribe G, Cleland RE, 1982: Active, irreversible accumulation of extreme levels of H_2SO_4 in the brown alga, *Desmarestia*.- Plant Physiology 70, 771-774

Mtolera M, Collén J, Pedersén M, Ekdahl A, Abrahamsson K, Semesi AK, 1996: Stress-induced production of volatile halogenated compounds in *Eucheuma denticulatum* (Rhodophyta) during stress caused by elevated pH and high light intensities.- European Journal of Phycology 31, 89-95

Müller DG, Westermeier R, Morales J, Garcia Reina G, del Campo E, Correa JA, Rometsch E, 2000: Massive prevalence of viral DNA in *Ectocarpus* (Phaeophyceae, Ectocarpales) from two habitats in the North Atlantic and South Pacific.- Botanica Marina 43, 157-159

Müller DG, Kapp M, Knippers R, 1998: Viruses in marine brown algae.- Advances in Virus Research 50, 49-67

Mur LAJ, Kenton P, Draper J, 1997: Something in the air: volatile signals in plant defence.- Trends in Microbiology 5, 297-300

Pedersén M, Collén J, Abrahamsson K, Ekdahl A, 1996: Production of halocarbons from seaweeds: an oxidative stress reaction?- Scientia Marina 60 (Suppl.1), 257-263

Pellegrini M, Pellegrini L, 1982: Some observations on relationships between bacteria and a brown alga. Biology of the Cell 43, 195-200

Penot M, 1974: Ionic exchange between tissues of *Ascophyllum nodosum* (L.) LE JOLIS and *Polysiphonia lanosa* (L.) TANDY.- Zeitschrift für Pflanzenphysiologie 73, 125-131

Peters AF, Burkhardt E, 1998: Systematic position of the kelp endophyte *Laminarionema elsbetiae* (Ectocarpales *sensu lato*, Phaeophyceae) inferred from nuclear ribosomal DNA sequences.- Phycologia 37, 114-120

Peters AF, van Oppen MJH, Wiencke C, Stam WT, Olsen JL, 1997: Phylogeny and historical ecology of the Desmarestiaceae (Phaeophyceae) support a southern hemisphere origin.- Journal of Phycology 33, 294-309

Peters AF, Schaffelke B, 1996: *Streblonema* (Ectocarpales, Phaeophyceae) infection in the kelp *Laminaria saccharina* (Laminariales, Phaeophyceae) in the western Baltic.- Hydrobiologia 326/327, 111-117

Pettit GR, 1996: Progress in the discovery of biosynthetic anticancer drugs.- Journal of Natural Products 59, 812-821

Phinney BO, 1983: The history of gibberellins.- In: The Biochemistry and Physiology of Gibberellins. Vol 1, Crozier A (ed.), Praeger, New York, 19-52

Prange A, Arzberger I, Engemann C, Modrow H, Schumann O, Trüper HG, Steudel R, Dahl C, Hormes J, 1999: In situ analysis of sulfur in the sulfur globules of phototrophic sulfur bacteria by X-ray absorption near edge spectroscopy.- Biochimica et Biophysica Acta 1428, 446-454

Pringle CR, 1998: Virus taxonomy San Diego.- Archives of Virology 143, 1449-1459

Real LA, 1996: Special feature: Disease ecology.- Ecology 77, 689

Reisser W, 1993: Viruses and virus-like particles of freshwater and marine eukaryotic algae – a review.- Archiv für Protistenkunde 143, 257-265

Reymond P, 2001: DNA microarrays and plant defence.- Plant Physiology & Biochemistry 39, 313-321

Ricker RW, 1987: Taxonomy and biogeography of Macquarie Island seaweeds.- British Museum Natural History, London

Royer-Pokora B, Kunkel LM, Monaco AP, Goff SC, Newburger PE, Baehner RL, Cole FC, Curnutte JT, Orkin SH, 1986: Cloning the gene for the inherited disorder chronic granulomatous-disease on the basis of its chromosomal location.- Nature 322, 32-38

Saunders GW, Druehl LD, 1993: Revision of the kelp family Alariaceae and the taxonomic affinities of *Lessoniopsis* Reinke (Laminariales, Phaeophyta).- Hydrobiologia 260/261, 689-697

Saunders GW, Kraft GT, 1995: The phylogenetic affinities of *Notheia anomala* (Fucales, Phaeophyceae) as determined from partial small-subunit rRNA gene sequences.- Phycologia 34, 383-389

Sawabe T, Tanaka R, Iqbal MM, Tajima K, Ezura Y, Ivanova EP, Christen R, 2000: Assignment of *Alteromonas elyakovii* KMM 162^T and five strains from spot-wounded fronds of *Laminaria japonica* to *Pseudoalteromonas elyakovii* comb. nov. and the extended description of the species.- International Journal of Systematic and Evolutionary Microbiology 50, 265-271

Sawabe T, Makino H, Tatsumi M, Nakano K, Tajima K, Iqbal MM, Yumoto I, Ezura Y, Christen R, 1998: *Pseudoalteromonas bacteriolytica* sp. nov., a marine bacterium that is the causative agent of red spot disease in *Laminaria japonica*.- International Journal of Systematic Bacteriology 48, 769-774

Sawabe T, Ezura Y, 1996: Regeneration from *Laminaria japonica* Areschoug (Laminariales, Phaeophyceae) protoplasts isolated with bacterial alginase.- Plant Cell Reports 15, 892-895

Schnitzler I, Boland W, Hay ME, 1998: Organic sulfur compounds from *Dictyopteris* spp. deter feeding by an herbivorous amphipod (*Ampithoe longimana*) but not by an herbivorous sea urchin (*Arbacia punctulata*).- Journal of Chemical Ecology 24, 1715-1732

Sengco MR, Bräutigam M, Kapp M, Müller DG, 1996: Detection of virus DNA in *Ectocarpus siliculosus* and *E. fasciculatus* (Phaeophyceae) from various geographic areas.- European Journal of Phycology 31, 73-78

Sparrow FK Jr., 1960: Aquatic Phycomycetes. - 2nd revised edition, The University of Michigan Press, Ann Arbor.

Stache-Crain B, Müller DG, Goff LJ, 1997: Molecular systematics of *Ectocarpus* and *Kuckuckia* (Ectocarpales, Phaeophyceae) inferred from phylogenetic analysis of nuclear- and plastid-encoded DNA sequences.- Journal of Phycology 33, 152-168

Tan IH, Druehl LD, 1994: A molecular analysis of *Analipus* and *Ralfsia* (Phaeophyceae) suggests the order of Ectocarpales is polyphyletic.- Journal of Phycology 30, 721-729

Theiler R, Carter Cook J, Hager LP, 1978: Halohydrocarbon synthesis by bromoperoxidase.- Science 202, 1094-1096

Torres MA, Onouchi H, Hamada S, Machida C, Hammond-Kosack KE, JDG Jones, 1998: Six *Arabidopsis thaliana* homologues of the human respiratory burst oxidase (*gp91*^{phox}).- Plant Journal 14, 365-370

Toth GB, Pavia H, 2000: Water-borne cues induce chemical defense in a marine alga (*Ascophyllum nodosum*).- Proceedings of the National Academy of Sciences of the USA 97, 14418-14420

Van Etten JL, Lane LC, Meints RH, 1991: Viruses and viruslike particles of eukaryotic algae.- Microbiol. Rev. 55, 586-620

Venturi S, Donati FM, Venturi A, Venturi M, 2000: Environmental iodine deficiency: a challenge to the evolution of terrestrial life?- Thyroid 10, 727-729

Vilter H, Glombitza KW, Grawe A, 1983: Peroxidases from Phaeophyceae I: extraction and detection of the peroxidases.- Botanica Marina 26, 331-340

Weinberger F, Friedlander M, 2000: Response of *Gracilaria conferta* (Rhodophyta) to oligoagars results in defense against agar-degrading epiphytes.- Journal of Phycology 36, 1079-1086

Weiss SJ, Test ST, Eckmann CM, Roos D, Regiani S, 1986: Brominating oxidants generated by human eosinophils.- Science 234, 200-203

Wever R, Tromp MGM, Krenn BE, Marjani A, Van Tol M, 1991: Brominating activity of the seaweed *Ascophyllum nodosum*: impact on the biosphere.- Environmental Science & Technology 25, 446-449

Wilce RT, Schneider CW, Quinlan AV, van den Bosch K, 1982: The life history and morphology of free-living *Pilayella littoralis* (L.) Kjellman (Ectocarpaceae, Ectocarpales) in Nahant Bay, Massachusetts.- Phycologia 21, 336-354

Wolfe JM, 1988: The chemical and community ecology of macroalgae in eight Rhode Island tidepools. Ph.D. dissertation, University of Rhode Island, Kingston

Wuosmaa AM, Hager LP, 1990: Methyl chloride transferase: A carbocation route for biosynthesis of halometabolites.- Science 249, 160-162

Yumoto I, Ezura Y, Kimura T, 1989: Distribution of the *Alteromonas* sp., the causative agent of red-spots on the culture bed of Makonbu *Laminaria japonica*, in the coastal area of Funka Bay.- Nippon Suisan Gakkaishi 55, 453-462

Summary

Associations of marine algae with symbiotic or parasitic microorganisms are ubiquitous phenomena known for a long time. However, there is an almost complete lack of knowledge on details of such interactions. The intention of this study is to use the potentials of modern biological and biochemical techniques in order to analyze the reaction of brown algal hosts to the attack by pathogens and epibionts.

A 3-year field study at different localities on the European Atlantic coast revealed that *Pylaiella littoralis* populations were subject to massive epidemics of the parasites *Eurychasma dicksonii*, *Chytridium polysiphoniae* and *Anisolpidium rosenvingei*. Laboratory cultures were used to investigate the association of *Eurychasma* and *Chytridium* with brown algal hosts from different taxonomic groups: *Eurychasma* has a much broader host range than *Chytridium*, comprising members of all brown algal orders investigated, and it tolerates a wider range of temperatures than the latter. Phylogenetic studies based upon 18 S rRNA genes revealed that *Eurychasma dicksonii* belongs to the Oomycota, branching at an ancestral position between terrestrial plant pathogens and free-living members of the marine heterotrophic picoplankton. *Chytridium polysiphoniae*, a fungus, appears more closely related to the genera *Rhizophydium* and *Spizellomyces* (Chytridiomycota) than to other known *Chytridium* species. *Chytridium* produces chitin, whilst *Eurychasma* does not. Early stages of *Eurychasma* infection have a rather modest effect on host physiology. The photosynthetic capacity is enhanced, suggesting a temporary stimulation of host metabolism for hypertrophic growth. *Chytridium*, in contrast, has an immediate detrimental effect on host photosynthesis, which breaks down once a cell is infected, leading to the rapid death of infected cells.

Laboratory cultures of kelp sporophytes (*Laminaria digitata*, *Macrocystis pyrifera*) were used to study biochemical responses of hosts to pathogen attack. Oligoalginates are recognized by *L. digitata* sporophytes as endogenous elicitors, triggering a rapid oxidative burst, an efflux of iodine, and a resistance to attack by a pathogenic endophyte (*Laminariocolax tomentosoides*). Furthermore, the oxidative burst also seems to convey natural resistance against alginolytic, bacterial epibionts and non-host resistance against *Eurychasma dicksonii*. Signal transduction following oligosaccharide recognition includes K^+, Cl^- and Ca^{2+} channels, protein phosphorylation and phospholipase A_2. The capacity of performing an oxidative burst appears as a characteristic feature of the alginate-rich, morphologically complex sporophytes of the Laminariales, whilst the vast majority of other brown algae do not show this response.

Laminaria digitata accumulates iodine up to several per cent of its dry weight by means of a peroxide-consuming, haloperoxidase-dependent, apoplastic mechanism. Upon oxidative stress, accumulated iodine can be mobilized, producing an efflux of various inorganic and organic iodinated compounds, some of which have potentially high anti-microbial toxicity. It is suggested that iodine accumulation has both anti-oxidant and defense functions.

Various lipid derivatives such as lipoteichoic acids and lipopolysaccharides from various bacteria, polyunsaturated fatty acids and methyl jasmonate were found to be potent triggers of the oxidative burst in kelps. Exogenous addition of prostaglandin A_2 triggered the strongest response ever observed, which might suggest that it has a key function in signal transduction following elicitation.

The results of this study confirm the existence of inducible defense mechanisms against pathogen attack in marine brown algae with some similarities to strategies in higher plants. The extent of the involvement of halogen metabolism, however, seems to be unique to the marine habitat.

Résumé

L'association d'algues marines avec des microorganismes symbiotiques ou parasites est un phénomène ubiquitaire connu depuis longtemps. Cependant, il n'existe que très peu de données relatives à ces interactions. L'objectif de ce travail est d'exploiter les possibilités des techniques modernes de biologie et biochimie afin d'analyser les réactions d'algues brunes à l'attaque de pathogènes ou de microorganismes épiphytes.

Une étude de terrain de trois ans menée en plusieurs endroits de la côte atlantique européenne a montré que les populations de *Pylaiella littoralis* sont l'objet d'épidémies massives dues à *Eurychasma dicksonii*, *Chytridium polysiphoniae* et *Anisolpidium rosenvingei*. L'utilisation de cultures clonales unialgales nous a permis d'étudier l'interaction d'*Eurychasma* et *Chytridium* avec différents groupes taxonomiques d'algues brunes : *Eurychasma* a un spectre d'hôtes bien plus large que *Chytridium*, comprenant des membres de tous les ordres d'algues brunes testés. Des études phylogénétiques basées sur le séquençage de leur ARNr 18S montrent qu'*Eurychasma dicksonii* est une espèce appartenant aux oomycètes, et qui est apparue tôt dans la lignée avant l'émergence des pathogènes de plantes terrestres. Certaines espèces de picoplankton hétérotrophes, qui semblent être des oomycètes aussi, auraient émergé encore plus tôt qu'*Eurychasma*.

Le champignon *Chytridium polysiphoniae* apparaît quant à lui plus apparenté aux genres *Rhizophydium* et *Spizellomyces* (Chytridiomycota) qu'aux autres espèces du genre *Chytridium*. *Eurychasma*, à l'inverse de *Chytridium*, ne produit pas de chitine. Aux stades précoces de l'infection, il a un effet plutôt minime sur la physiologie de son hôte, augmentant sa capacité photosynthétique, ce qui suggère une stimulation temporaire du métabolisme favorisant l'hypertrophie des cellules infectées. Au contraire, *Chytridium* a un effet délétère immédiat sur la photosynthèse de son hôte, qui chute rapidement à partir du moment où une cellule est infectée, ce qui contribue à la mort rapide de celle-ci.

Des cultures clonales de sporophytes de *Laminaria digitata* et *Macrocystis pyrifera* ont permis d'étudier la biochimie des réponses de l'hôte suite à l'attaque d'un pathogène. *L. digitata* reconnaît les oligoalginates comme des éliciteurs endogènes, ce qui provoque un burst oxydatif rapide, un efflux d'iode et induit une résistance à l'attaque par un pathogène endophyte (*Laminariocolax tomentosoides*). De plus, nos résultats suggèrent que le burst oxydatif intervient dans la mise en place d'une résistance naturelle à l'encontre de bactéries alginolytiques épiphytes et une résistance non-hôte vis-à-vis d'*Eurychasma dicksonii*. La voie de transduction du signal en aval de la reconnaissance des oligosaccharides implique des canaux K^+, Cl^-, et Ca^{2+}, des activités kinase et phospholipase A_2. La capacité à induire ce burst oxydatif apparaît comme une caractéristique des sporophytes de Laminariales au thalle morphologiquement complexe et riche en alginates, alors que la majorité des autres ordres d'algues brunes ne présentent pas une telle réactivité.

Laminaria digitata accumule jusqu'à 7 % de son poids sec en iode grâce à un mécanisme apoplastique consommateur de peroxyde et dépendant d'une activité haloperoxydase. Au cours d'un stress oxydant, l'iode accumulé peut être mobilisé, ce qui conduit à un efflux de divers composés iodés, organiques et inorganiques, dont certains ont potentiellement une forte activité antimicrobienne. L'accumulation d'iode aurait donc à la fois des fonctions de défense et de protection contre le stress oxydant.

Divers dérivés lipidiques tels que les acides lipotéichoïques, les lipopolysaccharides de différentes bactéries, des acides gras polyinsaturés ainsi que le méthyl-jasmonate se sont révélés être de puissants inducteurs du burst oxydatif chez les Laminariales. L'ajout exogène de prostaglandine A_2 provoque la réponse la plus intense jamais enregistrée, ce qui pourrait suggérer un rôle clé de ce composé dans la transduction du signal suite à l'élicitation.

Les résultats de cette étude confirment l'existence chez les algues brunes de mécanismes de défense inductibles par l'attaque de pathogènes, qui possèdent des ressemblances avec les stratégies connues chez les plantes supérieures. Cependant, l'ampleur de l'implication du métabolisme halogéné semble être une caractéristique unique liée à l'habitat marin.

Zusammenfassung

Wechselbeziehungen von marinen Algen mit symbiontischen und parasitischen Mikroorganismen sind weit verbreitet und seit längerem bekannt. In den meisten Fällen sind jedoch die Details derartiger Interaktionen unklar. Ziel dieser Studie ist, mit modernen biologischen und biochemischen Methoden die Reaktionen von Braunalgenwirten auf Befall durch Pathogene und Epibionten zu untersuchen.

In einer dreijährigen Freilanduntersuchung an verschiedenen Stellen der europäischen Atlantikküste wurden massive Epidemien von *Eurychasma dicksonii*, *Chytridium polysiphoniae* und *Anisolpidium rosenvingei* in Populationen von *Pylaiella littoralis* beobachtet. Mit Hilfe von Laborkulturen wurde die Interaktion von *Eurychasma* und *Chytridium* mit Angehörigen verschiedener Vertreter der Braunalgen untersucht. Es zeigte sich, daß das Wirtsspektrum von *Eurychasma* deutlich breiter als das von *Chytridium* ist. Die Zellwände von Chytridium enthalten Chitin, die von *Eurychasma* dagegen nicht. Phylogenetische Untersuchungen auf der Basis der 18 S rRNA-Gene zeigten, daß *Eurychasma* zu den Oomycota gehört. Er erscheint auf einem isolierten Ast zwischen terrestrischen Pflanzenpathogenen und frei lebenden Angehörigen des heterotrophen Picoplanktons. *Chytridium polysiphoniae* dagegen ist ein echter Pilz, der näher mit den Gattungen *Rhizophydium* und *Spizellomyces* (Chytridiomycota) verwandt ist als mit anderen Arten der Gattung *Chytridium*. Frühe Infektionsstadien von *Eurychasma* beeinträchtigen die Funktion der befallenen Wirtszellen nur schwach. Ihre Photosyntheseleistung wird sogar gesteigert, was eine vorübergehende Stimulierung des Wirtsstoffwechsels für das hypertrophe Wachstum andeutet. Dagegen hat *Chytridium* einen sofortigen, schädlichen Effekt auf die Wirtsphotosynthese, die nach Infektion einer Zelle schnell zusammenbricht und zum Absterben infizierter Zellen beiträgt.

Sporophyten-Laborkulturen von *Laminaria digitata* und *Macrocystis pyrifera* dienten der biochemischen Untersuchung von Abwehrreaktionen auf Pathogenbefall. Oligoalginate werden von *L. digitata* als endogene Elicitoren erkannt und induzieren einen schnellen *oxidative burst* und eine Resistenz gegen den Befall durch den pathogenen Endophyten *Laminariocolax tomentosoides*. Darüber hinaus trägt der *oxidative burst* zur natürlichen Resistenz gegenüber Alginat-abbauenden, bakteriellen Epibionten und *Eurychasma dicksonii* bei. Komponenten der Signalübertragung nach der Erkennung von Oligosaccharid-Elicitoren sind Kalium-, Chlorid- und Calcium-Kanäle, Proteinphosphorylierung und Phospholipase A_2. Der *oxidative burst* wurde als Besonderheit der Sporophyten von Laminariales gefunden, während er bei den meisten anderen Braunalgenordnungen und einfacheren Thallusformen zu fehlen scheint.

Laminaria digitata reichert Iod bis zu mehreren Prozent ihrer Trockenmasse an. An der Aufnahme ist ein apoplastischer, Haloperoxidase-abhängiger Mechanismus beteiligt. Unter oxidativem Stress, insbesondere bei einem *oxidative burst*, kann das gespeicherte Iod mobilisiert werden, wobei verschiedene anorganische und organische Iod-Species ausströmen, von denen einige hohe antimikrobielle Toxizität aufweisen. Die Befunde meiner Untersuchungen deuten darauf hin, daß die Anreicherung von Iod sowohl der Redox-Regulation als auch dem Abwehrstoffwechsel dient.

Verschiedene Lipid-Derivate, darunter Lipoteichonsäuren und Lipopolysaccharide aus diversen Bakterien, Linolen- und Arachidonsäure sowie Methyljasmonat sind starke Aktivatoren des *oxidative burst* in Braunalgen. Die exogene Zugabe von Prostaglandin A_2 löste die stärksten Reaktionen aus, die im Rahmen dieser Studie beobachtet wurden, was darauf hindeutet, daß sie eine Schlüsselfunktion bei der Signalübertragung nach Elicitation hat.

Die Ergebnisse dieser Studie belegen die Existenz induzierbarer Abwehrmechanismen gegen Pathogenbefall in marinen Braunalgen, wobei gewisse Ähnlichkeiten zu höheren Pflanzen bestehen. Das Ausmaß der Beteiligung des Halogenstoffwechsels an der Pathogen-Abwehr dagegen scheint eine Besonderheit des marinen Lebensraums zu sein.

Abgrenzung der Eigenleistung

Das Rahmenkonzept der Arbeit wurde von mir gemeinsam mit den beiden Betreuern meiner Promotion, Prof. Dr. Dieter G. Müller (Universität Konstanz) und Dr. Bernard Kloareg (CNRS, Roscoff) aufgestellt, wobei Dr. Kloareg überwiegend den biochemisch-physiologischen und Prof. Müller den biologisch-ökologischen Teil der Arbeit betreute.

Die Kapitel 1-2, 4-7 sowie 9-11 wurden von mir selbst federführend verfaßt; mit Ausnahme von Kapitel 1 entsprechen sie teilweise erschienenen (2, 6, 7), eingereichten (4, 9) oder in Vorbereitung befindlichen (5, 10) Manuskripten von Veröffentlichungen, deren Erstautor ich bin. Die beschriebenen Versuche wurden zum Großteil von mir durchgeführt; die Coautoren waren an experimentellen Teilprojekten beteiligt. Ausnahmen stellen die Kapitel 3 und 8 dar: Die Infektionsversuche von Kapitel 3 wurden von Prof. Müller in Konstanz durchgeführt, während ich in Roscoff Freiland- und Laborarbeiten weiterführte. Wegen seines verhältnismäßig großen Anteils an diesem Teilprojekt wurde das Manuskript weitgehend von ihm als Erstautor verfaßt. Kapitel 8 ist das Ergebnis eines Gemeinschaftsprojektes mit Dr. Gillian Malin an der University of East Anglia, angeregt durch Dr. Bernard Kloareg: Nachdem ich in Roscoff in Isotopenversuchen beobachtet hatte, daß der *oxidative burst* zu einer Mobilisierung des Iodspeichers in Laminarien führt, hatten wir eine Zusammenarbeit aufgenommen, um mit analytisch-chemischen Methoden dem Phänomen auf den Grund zu gehen. Die Isotopenversuche wurden komplett von mir selbst durchgeführt. Die analytischen Befunde sind Teil von zwei Versuchsreihen in Norwich sowie an der Mace Head Atmospheric Research Station (Co. Galway, Irland). Konzeption und Durchführung der Inkubationen von Algen erfolgten durch Dr. Malin und mich, während die Analysen von folgenden Kollegen durchgeführt wurden: Dr. Carpenter (halogenierte Kohlenwasserstoffe), Dr. Baker (anorganische Iod-Species) und Dr. Broadgate (Kohlenwasserstoffe).

Curriculum Vitae

Frithjof Christian Küpper

Personal data & education

1972-05-25	born in Marl (Westfalen), Germany
1978 to 1991	elementary and secondary school in Marl
1988-09 to 1988-12	De la Salle College, Dublin, Eire / Ireland (*German-Irish Cultural Exchange*)
1991-06-13	graduation (Abitur; mark: 1.0 / "sehr gut") from secondary school (Geschwister-Scholl-Gymnasium, Marl, Germany)
1991-07-01 to 1992-09-30	civil service in the nature and environment conservation office in Recklinghausen (Westfalen)
1992-10 to 1997-11	Studies of biology at the University of Konstanz
1994-07-21	first undergraduate exam (Vordiplom)
1997-03	oral part of the graduate exam (Diplom), respectively ("sehr gut")
1997-04 to 1997-11	experimental work for the final exam at the Station Biologique de Roscoff under supervision of Professor Dr. D.G. Müller (Plant Physiology, University of Konstanz): "Host-pathogen interactions in marine brown algae"
1997-26-11	Diplombiologe (mark: 1.0, "sehr gut")
1998-03 to 2001	Doctorate studies at CNRS-Roscoff / Université de Paris-Sud XI and University of Konstanz: "Biology and biochemistry of host-pathogen interactions in marine brown algae"
2000-05	EU-ESF Advanced Training Course "Chemistry of Metals in Biological Systems" (Louvain-la-Neuve, Belgium)

Professional experience outside the regular curriculum of studies

1993-03-01 to 1993-03-31	trainee in the Dept. of Microbiology & Toxicology of Hüls AG, Marl, Germany
1994-03-01 to 1994-04-15	trainee at the National Centre for Marine Research, Agios Kosmas, Athens, Greece (Dept. of Phytoplankton and Phytobenthos)
1995-09 to 1996-07	trainee at the Station Biologique de Roscoff, France (CNRS, UPR 9042)
1998-04-11 to 1998-05-11	participation in a one-month expedition to French Polynesia searching new anti-cancer substances from marine organisms and rainforest plants (supervised by Prof. G.R. Pettit, Cancer Research Institute, Arizona State University, Tempe, Arizona, USA, and Dr. L. Meijer, CNRS, Roscoff, France)

Publications

(peer-reviewed articles in periodicals)

Küpper F.C., Kloareg B., Guern J., Potin P., 2001: Oligoguluronates elicit an oxidative burst in the brown algal kelp *Laminaria digitata*.- Plant Physiology 125, 278-91

Küpper H., Spiller M., Küpper F.C., 2000: Photometric method for the quantification of chlorophylls and their derivatives in complex mixtures: fitting with gauss-peak-spectra.- Analytical Biochemistry 286, 247-56

Carpenter L.J., Malin G., Küpper F.C., Liss P.S., 2000: Novel biogenic iodine-containing trihalomethanes and other short-lived halocarbons in the coastal East Atlantic.- Global Biogechemical Cycles 14 (4), 1191-1204

Müller D.G., Küpper F.C., Küpper H., 1999: Infection experiments reveal broad host ranges of *Eurychasma dicksonii* (Oomycota) and *Chytridium polysiphoniae* (Chytridiomycota), two eukaryotic parasites in marine brown algae (Phaeophyceae).- Phycological Research 47, 217-23

Küpper F.C., Müller D.G., 1999: Massive occurrence of the heterokont and fungal parasites *Anisolpidium*, *Eurychasma* and *Chytridium* in *Pylaiella littoralis* (Ectocarpales, Phaeophyceae).- Nova Hedwigia 69, 381-9

Küpper H., Küpper F.C., Spiller M., 1998: In situ detection of heavy metal substituted chlorophylls.- Photosynthesis Research 55, 123-33

Küpper F.C., Schweigert N., ArGall E., Legendre J.M., Vilter H., Kloareg B., 1998: Iodine uptake in Laminariales involves extracellular, haloperoxidase-mediated oxidation of iodide.- Planta 207, 163-71

Küpper H., Küpper F.C., Spiller M., 1996: Environmental relevance of heavy metal substituted chlorophylls using the example of submersed water plants. - Journal of Experimental Botany 47 (295), 259-266

(review article)

Potin P., Bouarab K., Küpper F.C., Kloareg B., 1999: Oligosaccharide recognition signals and defence reactions in marine plant-microbe interactions.- Current Opinion in Microbiology 2: 276-83

(miscellaneous articles)

Küpper F.C., 2001: New record of *Anisolpidium rosenvingei* (H.E. Petersen) Karling in Ireland.- Irish Naturalists' Journal, *in press*

Broadgate W., Carpenter L., Malin G., Thompson A., Küpper F., 2000: Non-methane hydrocarbons (NMHCs), dimethyl sulphide (DMS) and halocarbons.- In: O'Dowd C., Hämeri K. (ed.): New Particle Formation and Fate in the Coastal Environment.- Report Series in Aerosol Science No. 48, Finnish Association for Aerosol Research, Helsinki

Küpper H., Küpper F. , Spiller M., 1997: Schwermetallchlorophylle in vivo.- junge wissenschaft 45, 47-53

Küpper F. , Küpper H., Spiller M., 1996: Eine aggressive Wasserpflanze aus Australien und Neuseeland: *Crassula helmsii* (KIRK) COCKAYNE.- Floristische Rundbriefe 30 (1), 24-29

Küpper F. , Küpper H., Spiller M., 1995: *Crassula helmsii* (KIRK) COCKAYNE. Ein Neuankömmling in der europäischen Süßgewässerflora.- Naturkundliche Beiträge des DJN 30, 45-49

(book review)

Küpper F. , 1996: (review of) Molecular ecology and evolution. B. Schierwater (ed.), Birkhäuser, Basel, 1995.- junge wissenschaft 42, 75-76

(conference contributions)

Küpper F.C., Leblanc C., Colin C., Malin G., Kloareg B., Potin P., 2001: Pathogen recognition and defense mechanisms in Laminariales.- (*Lecture by F.C. Küpper at the International Seaweed Symposium, Cape Town, South Africa*)

Leblanc C., Küpper F., Potin P., Kloareg B., 1999: Defence mechanisms and halogen metabolism in brown algae.- *(Lecture by C. Leblanc at the European Phycological Congress, Montecatini Terme/Italy)*

Küpper F., Schweigert N., Legendre J.-M., Vilter H., Kloareg B., 1997: Iodine uptake involves extracellular haloperoxidase - mediated oxidation of iodide.- *(Lecture by B. Kloareg at the International Phycological Congress, Leiden/Netherlands)*

Schweigert N., Küpper F. , Morin P.P., Vilter H., Kloareg B., 1996: Iodine uptake in Laminariales involves extracellular oxidases.- *(Poster at the European Phycological Congress, Cologne/Germany)*

Küpper H., Küpper F. , Spiller M., 1994: Environmental relevance of heavy metal substituted chlorophylls using the example of submersed water plants. - (*Lecture at the Euroconference "Science and Technology, Lever of the New European Development", Oporto/Portugal*)

Foreign languages

English, modern Greek, French (all spoken fluently), Latin (Latinum 1988)

Grants and fellowships

February 1993-Oct. 1997	fellow of the Studienstiftung des deutschen Volkes (Bonn), supported during the studies in Konstanz, Agios Kosmas and Roscoff
September 1995	grant of the Hüls AG-Stiftung (Marl), supporting the stay at the Station Biologique de Roscoff 1995/96, additional coverage for the purchase of books and scientific equipment
April 1998	travel grant of the Boehringer Ingelheim Fonds (Stuttgart) for the participation in a one-month expedition to French Polynesia with Prof. G.R. Pettit (Arizona State University, Tempe, Arizona, USA)
July 1998-June 2001	Ph.D. fellowship from Studienstiftung des Deutschen Volkes (Bonn)

April 1999-April 2001	Marie Curie Research Training Grant (MAST III program) from the European Commission

Competitions and awards

February 1988	1st prize in the section of biology at the regional competition *Jugend forscht / Schüler experimentieren* (German young scientists' competition) with a contribution about "Plant pigments and autumnal foliages" (together with my younger brother Hendrik und Martin Spiller)
December 1991	1st monthly prize in the nation-wide competition *Jugend schützt Umwelt / Sven-Simon-Preis* ("Youth protects the environment / Sven Simon Award") with a contribution about "Conservation and observations of hedgehogs in a wildlife garden" (together with Hendrik Küpper)
February and March 1992	1st prizes at the regional and country-wide competition of *Jugend forscht* in North-Rhine Westfalia in the sections of chemistry and biology, respectively, as well as the special awards of the ministry of the environment of North-Rhine Westfalia at both competitions
1992-05-11	1st prize at the German national competition *Jugend forscht* in the section of biology (Duisburg, Germany)
1992-09-23	1st prize at the *4th European Community Contest for Young Scientists* (Seville, Spain)
1992-11-22	1st prize at the 3rd competition *Young Europeans' Environmental Research/ Europas Jugend forscht für die Umwelt* (Essen, Germany)

At all competitions in 1992 the following contribution was presented (together with H. Küpper and M. Spiller): "Environmental relevance of heavy metal substituted chlorophylls using the example of submersed water plants"

2001-02-02	3rd prize: University of British Columbia Award for the Best Student Paper Contribution (University of Cape Town, South Africa)

Acknowledgements

I am grateful to Prof. Dr. Dieter G. Müller, Universität Konstanz, who recognized and supported my interests in marine plants several years ago and who paved my way to Roscoff. While in Roscoff, he supported my studies in all aspects, always remaining in intensive exchange of ideas despite the long distance from home. In Konstanz, he supported me with his extensive algal culturing facilities of the highest standards, without which the infection experiments of my study would have been unimaginable.

In Roscoff, I would first of all like to thank Dr. Bernard Kloareg for the hospitable reception in his lab, the generous support with research facilities and in finding suitable funding for an unusual project, the frequent, intensive discussions, and for his support to send me on research trips to the Shetland Islands, Norwich, Ireland, and more recently, to the Seaweed Symposium in Cape Town. More significantly, he has allowed me to take care of a student called Claire...: Trugarez, Bernez.

My experimental work in Roscoff would have been unimaginable without the permanent and patient support from Dr. Philippe Potin. His family contributed a spiny aspect to my life, Kiki, the first ever hedgehog to have lived in the Roscoff marine lab, who remains unforgettable to many members of the institution.

I am deeply indebted to Dr. Susan Loiseaux-de Goer for our many stimulating, encouraging and amusing discussions, for her initiative to undertake a phylogenetic study of my algal pathogens. On the human side, she considerably brightened up my life with her genuine sense of humour, a rare composition of British and Gaulish ingredients extracted with *uisge beatha*, and the outstanding feasts in her home, unforgettable for everybody involved.

The arrival of Dr. Jean-Pierre Salaün in our research unit, coinciding with the second half of my Ph.D., was a rare luck, enriching my work on algal defense by a topic of a growing discipline, lipid signalling. I should not forget mentioning his and his wife's hospitality, marvellous wines and dinners in their home before their moving from Strasbourg to Brittany.

Centered in Konstanz and Roscoff, this work still implied on a number of external collaborations. I would like to thank Dr. Akira F. Peters (Institut für Meereskunde, Kiel) for his crucial role in infection studies of brown algae, Dr. Hans Vilter (Trier) for his support in purifying haloperoxidase, Dr. Lucy Carpenter, Dr. Wendy Broadgate, Prof. Dr. Peter S. Liss and Dr. Gillian Malin (all at the University of East Anglia, Norwich) for the fruitful project on stress-related trace gas emissions by algae, Dr. Jean-M. Legendre (Centre Hospitalier Universitaire, Brest) for radioactive tracer studies with iodine, and Dr. Ivan Setlik (Academy of Sciences of the Czech Republic, Trebon) and my brother Hendrik (Trebon and Konstanz) for their hospitable reception in Bohemia, enabling a pioneering study using fluorescence kinetic microscopy.

In the laboratories in Roscoff and Konstanz, I would still like to mention Dr. Ingo Maier (including for the daily delivery of *Brez'n*), Dr. Nicolas Delaroque, Klaus Sulger, Dr. Volker Hellmann, Dr. Sabine Genicot, Dr. Florian Weinberger, Dr. Catherine Leblanc, Dr. Jean-Francois Lennon, Sylvie Rousvoal, Kamal Bouarab, Sandrine Boulben, Dr. Fadi Adas, Dr. Jean-Baptiste Marchand, Dr. Laure Guillou, for their pleasant company and co-operative attitudes in everyday lab work. I would like to extend the same gratitude to Dr. Aldo Asensi for his support before his retirement in Roscoff, and his hospitality in Paris thereafter.

The University of Konstanz was an important point of anchorage throughout my Ph.D., and I hope it will remain it in the future. In the Department of Biology, I am grateful to Prof. Dr. Peter M.H. Kroneck, Prof. Dr. Alasdair M. Cook, Prof. Dr. Werner Rathmayer, Prof. Dr. Günther Stark and to Prof. Dr. Bernhard Schink, for idealistically fostering my activities, by their availability for numerous useful discussions, and support in finding grants. At the Université de Paris-Sud XI / Orsay, I am indebted to Prof. Dr. Thierry Langin for his patient backing of my plan of a bi-national doctorate (thèse en cotutelle), which started with a late-night meeting in an RER suburban train station in Bourg-la-Reine, in the *banlieue* of Paris.

I am indebted to old friends, Dr. Roy M. Pinkerton (University of Edinburgh), Michael Doyle (Co-Operation Ireland, Dublin), Martin Hesselsoe (Royal Veterinary and Agricultural University, Copenhagen) for retrieval of historic literature; to Tewes Tralau (University of Konstanz), for support whenever computing problems occurred; and to Eva-Maria Boneberg (University of Konstanz) and Giorgos K. Tofaris (Trinity College, Cambridge) for stimulating discussions about parallels between inflammatory reactions in humans and events around the oxidative burst in brown algae.

The best scientific ideas would have to remain hypothetical if there was nobody to pay for their realization. I gratefully acknowledge Studienstiftung des Deutschen Volkes (Bonn, Germany), Hüls AG Stiftung (Marl, Germany), Boehringer Ingelheim Fonds (Stuttgart, Germany) and the Directorate General XII of the European Commission for the generous funding and outstanding idealistic initiatives provided to foster my studies and to exchange views with established scientists and other scholars.

I wish to close by expressing my gratitude towards my family, my parents Eva-Maria and Friedrich-Wilhelm, to Claire and to Hendrik, who have made this work possible in the most diverse ways, and who often adopted their own lives correspondingly.

I would like to dedicate this work to them.